SpringerBriefs in Energy

For further volumes:
http://www.springer.com/series/8903

Zhebo Chen · Huyen N. Dinh
Eric Miller

Photoelectrochemical Water Splitting

Standards, Experimental Methods, and Protocols

 Springer

Zhebo Chen
Department of Chemical Engineering
Stanford University
Stanford, CA
USA

Eric Miller
Fuel Cell Technologies
U.S. Department of Energy
Washington, DC
USA

Huyen N. Dinh
National Renewable Energy Laboratory
Hydrogen Technologies
  and Systems Center
Golden, CO
USA

ISSN 2191-5520　　ISSN 2191-5539　(electronic)
ISBN 978-1-4614-8297-0　　ISBN 978-1-4614-8298-7　(eBook)
DOI 10.1007/978-1-4614-8298-7
Springer New York Heidelberg Dordrecht London

Library of Congress Control Number: 2013943566

Printed on acid-free paper

Springer is part of Springer Science+Business Media (www.springer.com)

Preface

The methods and definitions presented herein are the product of an effort supported by the U.S. Department of Energy (DOE) to form a consensus among a number of experienced researchers in the area of photoelectrochemical (PEC) hydrogen production from various DOE-supported laboratories (including national labs and academic institutions) and other international partners. An early result of this effort was the production of a consolidated version of the guide presented here, published in the form of a review paper in the Journal of Materials Research in 2010, [1] and is reprinted in part with permission in this book. The extended guidance in the present work aims to accelerate materials development by establishing standards for methods, definitions, and reporting protocols that will enable direct cross-comparison of materials' properties and performance metrics. The intent is to facilitate knowledge transfer on a global scale.

The authors who have contributed to the writing of this book are members of the PEC Standards Working Group, assembled by the Energy Efficiency and Renewable Energy's Fuel Cell Technologies Office at the U.S. DOE. These authors include:

- Zhebo Chen (Department of Chemical Engineering, Stanford University, Stanford, CA 94305-5025)
- Todd G. Deutsch (Hydrogen Technologies and Systems Center, National Renewable Energy Laboratory, Golden, Colorado 80401)
- Huyen N. Dinh (Hydrogen Technologies and Systems Center, National Renewable Energy Laboratory, Golden, Colorado 80401)
- Kazunari Domen (Department of Chemical System Engineering, University of Tokyo, 7-3-1 Hongo, Bunkyo-ku, Tokyo 113-8656, Japan)
- Keith Emery (National Center for Photovoltaics, National Renewable Energy Laboratory, Golden, Colorado 80401)
- Arnold J. Forman (Department of Chemistry and Biochemistry, University of California—Santa Barbara, Santa Barbara, CA 93106-5080)
- Nicolas Gaillard (Hawaii Natural Energy Institute, University of Hawaii at Manoa, Honolulu, HI 96822)
- Roxanne Garland (Fuel Cell Technologies Office, U.S. Department of Energy, Washington, D.C. 20585)

- Clemens Heske (Department of Chemistry, University of Nevada—Las Vegas, Las Vegas, NV 89154-4003)
- Thomas F. Jaramillo (Department of Chemical Engineering, Stanford University, Stanford CA 94305-5025)
- Alan Kleiman-Shwarsctein (Department of Chemical Engineering, University of California—Santa Barbara, Santa Barbara, CA 93106-5080)
- Eric Miller (Fuel Cell Technologies Office, U.S. Department of Energy, Washington, D.C. 20585)
- Kazuhiro Takanabe (Division of Physical Sciences and Engineering, King Abdullah University of Science and Technology (KAUST), Thuwall, Kingdom of Saudi Arabia)
- John Turner (Hydrogen Technologies and Systems Center, National Renewable Energy Laboratory, Golden, Colorado 80401)

The chapters written herein represent many years of collaborative discussion and review, and we hope that their content can enable new researchers in the field of PEC water splitting to rapidly gain traction in their own laboratories towards the development of high efficiency materials.

A number of international researchers participated in providing excellent feedback on the text written in this book, including many affiliated with the International Energy Agency's Hydrogen Implementing Agreement Task 26. In no particular order, we thank Grant Mathieson, Bruce Parkinson, Jennifer Leisch, Theanne Schiros, David Peterson, Ib Chorkendorff, Peter Vesborg, Kendra Kuhl, Blaise Pinaud, Julie Tuttle, Sarah Havig, Berc Kalanyan, Billie Abrams, Candace Chan, Nelson Kelly, Shiwei Lin, Nianqiang Wu, Shane Ardo, Nick Strandwitz, Lorna Jeffery Mingu, Daniel Schaadt, Marie Mayer, Ke Sun, Nikolaos Vlachopoulos, Qiang Huang, Juan Hodelin, Jian Jin, Anna Goldstein, Kevin Sivula, Kazuhiro Sayama, Isabell Thomann, Yue Tak Lai, Ilwhan Oh, and Sonia Juliana Calero.

Reference

1. Z. Chen, T.F. Jaramillo, T.G. Deutsch, A. Kleiman-Shwarsctein, A.J. Forman, N. Gaillard, R. Garland, K. Takanabe, C. Heske, M. Sunkara, E.W. McFarland, K. Domen, E.L. Miller, J.A. Turner, H.N. Dinh, J. Mater. Res. **25**(1), 3–16 (2010)

Contents

Chapter 1
Introduction

One of the most important technical problems facing humanity is the development of a long-term, sustainable energy economy [1]. Although we have the technology and sufficient coal reserves to provide the energy needed for centuries' worth of population growth and economic development, this strategy could come with catastrophic societal costs [2]. Scientific discovery and innovation will be vital to achieving environmentally sound and cost-effective solutions. Technologies are needed that will meet society's increase in annual global energy consumption from the 495 quadrillion British thermal units (Btu) per year we consumed in 2007 to the projected demand of 739 quadrillion Btu/yr by the year 2035 [3]. In terms of power consumption, this corresponds to an increase from 16.6 terawatts (TW) to 24.7 TW. Solar photoelectrochemical (PEC) hydrogen production is one of the promising technologies that could potentially provide a clean, cost-effective, and domestically produced energy carrier by taking advantage of the $\sim 120,000$ TW of radiation that continually strikes the earth's surface [4].

The concept of PEC water splitting for hydrogen production has been investigated for decades, with first demonstration in 1972 by Fujishima and Honda [5]. In 1998, Khaselev and Turner [6] demonstrated a PEC solar-to-hydrogen conversion efficiency of 12.4 %, highlighting the great potential for a PEC technology which combines the harvesting of solar energy and the electrolysis of water into a single device. Basically, when a PEC semiconductor device with the right set of properties is immersed in an aqueous electrolyte and irradiated with sunlight, the photon energy is converted to electrochemical energy, which can directly split water into hydrogen and oxygen (chemical energy). Thus intermittent solar energy is converted into an inherently more storable form of energy, that of chemical bonds.

PEC solar water splitting is a powerful, but complex process. For direct photoelectrochemical decomposition of water to occur efficiently and sustainably, several key criteria must be met simultaneously: the semiconductor system must generate sufficient voltage upon irradiation to split water, the bulk band gap must be small enough to absorb a significant portion of the solar spectrum, the band edge potentials at the surfaces must straddle the hydrogen and oxygen redox potentials, the system must exhibit long-term stability against corrosion in aqueous electrolytes, and finally, the charge transfer from the surface of the semiconductor

Z. Chen et al., *Photoelectrochemical Water Splitting*,
SpringerBriefs in Energy, DOI: 10.1007/978-1-4614-8298-7_1,
© The Author(s) 2013

to the solution must be facile to minimize energy losses due to kinetic overpotential and selective for the hydrogen evolution reaction (HER) and oxygen evolution reaction (OER). To date, no cost-effective materials system satisfies all of the technical requirements listed above for practical hydrogen production. While research and development is ongoing to discover materials with bulk and interfacial characteristics that meet these criteria, advances in material science and interfacial electrochemistry are still needed.

Many previously published books [7–9] and review articles [10–12] include excellent discussions of the fundamental principles of PEC. Still, a brief summary of the basic operations of PEC devices is instructive here. Figure 1.1a illustrates fundamental processes in a PEC device for the example of a two-electrode system containing a single absorber photoanode. Incoming photons ($h\nu$) generate electrons (e^-) and holes (h^+) with an efficiency labeled η_{e^-/h^+}. The photogenerated electrons and holes then separate and travel through the semiconductor in opposite directions; the efficiency associated with the charge transport process is labeled $\eta_{transport}$. The holes drive the OER at the surface of the semiconductor working electrode. Simultaneously, the electrons are driven to the rear ohmic contact and through an electrical connection to the surface of the counter electrode to drive the HER. The combined efficiency of the charge transfer at the solid–liquid interface for both electrons and holes is labeled $\eta_{interface}$. An important discussion of how these processes affect calculations of PEC device efficiencies is contained in Chapter "Efficiency Definitions in the Field of PEC": Figure 1.1a also depicts the minimum thermodynamic voltage for splitting water ($\Delta E^0 = 1.23$ V).

A more extensive illustration of the energetic requirements for a PEC water splitting device is shown in Fig. 1.1b. In addition to the thermodynamic requirement, there are overpotentials associated with driving the kinetics of the hydrogen evolution reaction (OP_{HER}) and oxygen evolution reaction (OP_{OER}) at the solid–liquid interface. Minimizing these overpotentials through the development of efficient catalysts for each half-reaction is a key step in making highly efficient water splitting devices.

Entropic losses associated with the photogenerated electrons and holes must also be considered [13, 14]. The actual driving force for water splitting is represented as a photovoltage (V_{ph}) which, as a result of losses (arising from factors such as spontaneous emission, incomplete light trapping, and non-radiative recombination) [13], is always less than the band gap of the semiconductor. Additional factors such as non-ideal band structure alignment can further reduce available photovoltage. The photovoltage in the semiconductor is the potential difference between the quasi-Fermi levels of electrons ($E_{F,n}$) and holes ($E_{F,p}$) under illumination, although the accuracy of this formalism, particularly in the vicinity of semiconductor-liquid interface, remains an active area of discussion within the PEC community. A more conservative requirement for unassisted water splitting is that the photovoltage (compensated for overpotential losses) must enable the quasi-Fermi levels under illumination to straddle the OER and HER redox potentials.

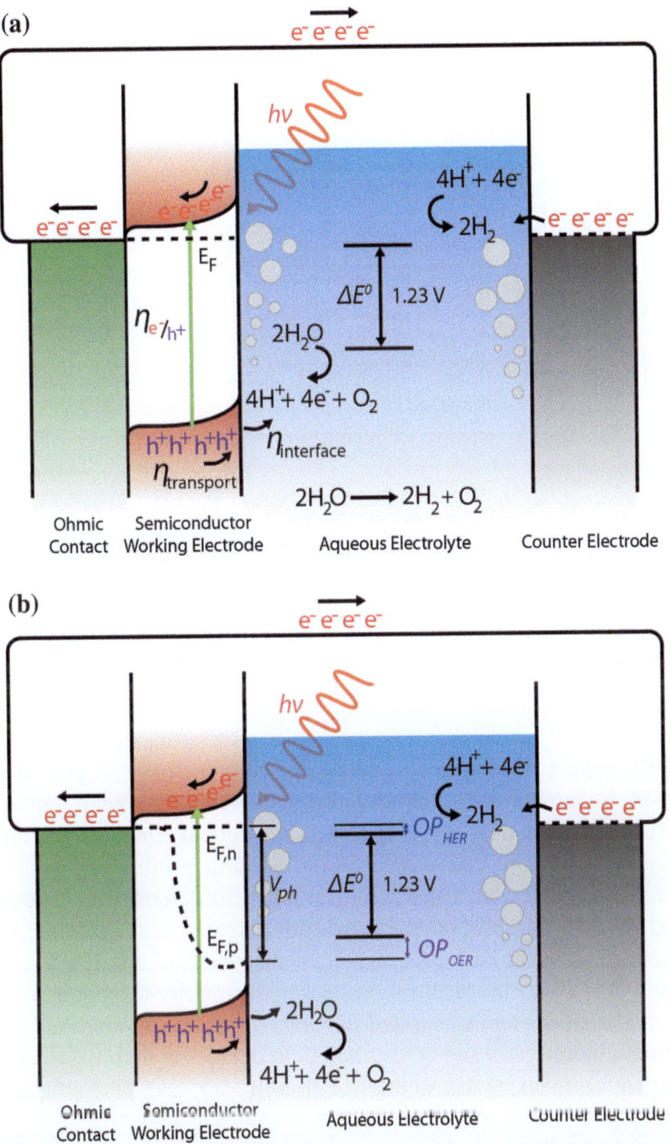

Fig. 1.1 Band structure of an n-type photoanode water splitting device, (**a**) Illustrating the various processes of photon irradiation, electron–hole pair formation, charge transport, and interfacial reactions, (**b**) Illustrating the energetic requirements associated with the minimum thermodynamic energy to split water, catalytic overpotentials for the HER and OER half-reactions, and photovoltage

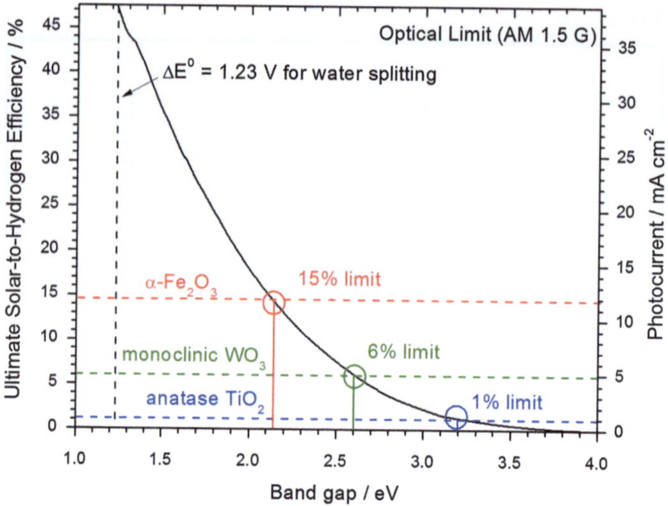

Fig. 1.2 Theoretical maximum solar-to-hydrogen (STH) conversion efficiency (*left axis*) and photocurrent (*right axis*) as a function of material band gap. The theoretical maximum STH plotted here only considers the first thermodynamic principle of energy conservation and is analogous to the ultimate efficiency for a p–n junction solar cell described by Shockley and Queisser [14]

As cited frequently, it follows that the semiconductor material band gap is a key factor in PEC solar-to-hydrogen (STH) conversion efficiency. A plot of maximum theoretical photocurrent and potential STH efficiency versus band gap is shown in Fig. 1.2. Although the plot only represents an optical limit (assuming all solar photons are captured and converted without additional efficiency loss), it is still illustrative of the important role band gap plays in determining theoretical efficiency, and serves to guide researchers toward materials that maximize absorption. It is important, however, to emphasize that the value of the band gap alone is insufficient to describe water splitting capability, primarily for the reasons stated in the previous paragraph and illustrated in Fig. 1.1b. If the photovoltage developed by the semiconductor band gap is insufficient to overcome thermodynamic barriers and overpotential losses, water splitting will not occur, and STH efficiency cannot be defined.

Despite the challenges, there are promising pathways for achieving the important goal of efficient, cost-effective PEC hydrogen production. For continued progress in overcoming the most important remaining scientific and engineering barriers, widely accepted standards in the characterization and reporting of PEC materials and devices are needed. This is vital to enable researchers from different groups to report in ways which allow for direct comparisons. It is expected that worldwide adoption of standardized methods for characterizing, reporting and screening PEC materials systems will result in more accurate and reliable information, and will facilitate the decision-making process for prioritizing research resources toward the most promising pathways. Quantifying technological barriers and bottlenecks through standardized methods will also prove to be invaluable in establishing research focus areas and new directions to address these challenges. A thorough and commonly-accepted understanding of the materials systems for PEC solar hydrogen production will certainly facilitate research and development, and advance the field toward developing devices that meet all of the PEC requirements.

The initial set of PEC techniques and protocols described in this work represent a first step toward standardization. Generally, they are not time-intensive to perform and require only basic electrochemical and optical equipment (e.g. potentiostat, UV-Visible spectrophotometer, tungsten/xenon light source, etc.). Still, these key techniques can provide critical information needed to guide researchers toward advanced materials systems for efficient PEC water splitting. As research and development progresses, the development of a comprehensive set of standard protocols will become increasingly important to the research community for providing guidance on what are acceptable and unacceptable practices in the characterization of PEC materials as well as the measurement and reporting of efficiencies and the stability of such devices. It is expected that the methods described in this publication will be refined and expanded in future work.

References

1. J.P. Holdren, Energy and sustainability. Science **315**, 737 (2007)
2. N.S. Lewis, D.G. Nocera, Powering the planet: chemical challenges in solar energy utilization. Proc. Natl. Acad. Sci. U. S. A. **104**, 20142 (2007)
3. L.E. Doman, K.A. Smith, L.D. Mayne, E.M. Yucel, J.L. Barden, A.M. Fawzi, P.D. Martin, V.V. Zaretskaya, M.L. Mellish, D.R. Kearney, B.T. Murphy, K.R. Vincent, P.M. Lindstrom, M.T. Leff, *International Energy Outlook 2010* (2010)
4. J. Stringer, L. Horton, *Basic Research Needs to Assure a Secure Energy Future* (2003)
5. A. Fujishima, K. Honda, Electrochemical photolysis of water at a semiconductor electrode. Nature **238**, 37–38 (1972)
6. O. Khaselev, J.A. Turner, A monolithic photovoltaic-photoelectrochemical device for hydrogen production via water splitting. Science **280**, 425–427 (1998)
7. M.D. Archer, A.J. Nozik, *Nanostructured and Photoelectrochemical Systems for Solar Photon Conversion*, vol. 3 (Imperial College Press, London, 2008)
8. R. Memming, *Semiconductor Electrochemistry* (Wiley-Vch, Weinheim, 2001)
9. L. Vayssieres, *On Solar Hydrogen and Nanotechnology* (Wiley, Chichester, 2009)
10. T. Bak, J. Nowotny, M. Rekas, C.C. Sorrell, Photo-electrochemical hydrogen generation from water using solar energy. Materials-related aspects. Int. J. Hydrog. Energy **27**, 991–1022 (2002)
11. A. Kudo, Y. Miseki, Heterogeneous photocatalyst materials for water splitting. Chem. Soc. Rev. **38**, 253–278 (2009)
12. M. Gratzel, Photoelectrochemical cells. Nature **414**, 338–344 (2001)
13. A. Polman, H.A. Atwater, Photonic design principles for ultrahigh-efficiency photovoltaics. Nat. Mater. **11**, 174–177 (2011)
14. W. Shockley, H.J. Queisser, Detailed balance limit of efficiency of p–n junction solar cells. J. Appl. Phys. **32**, 510–519 (1961)

Chapter 2
Efficiency Definitions in the Field of PEC

2.1 Overview of Efficiency Definitions

Overall solar-to-hydrogen (STH) efficiency is the most important parameter to characterize a PEC device. In fact, materials systems themselves are effectively defined by their highest-recorded STH efficiency; it is the single value by which all PEC devices can be reliably ranked against one another [1]. Unfortunately, published literature in the area of PEC sometimes contains confusing information regarding efficiency including invalid mathematical expressions for device efficiency, improper experimental methods for obtaining efficiency values, and/or wide-scale reporting of efficiencies other than STH without clear distinction. The first goal of this document is to establish proper definitions and mathematical expressions for device efficiencies. Among these definitions, we identify those that are acceptable for wide-scale benchmarking and reporting (for instance in the form of press releases to mainstream media) as well as those definitions which are helpful for their scientific value in material characterization and diagnostic testing (and suitable for journal publications). Later in this document, we overview the proper experimental procedures as well as common pitfalls that concern each type of efficiency measurement.

One of the reasons why so much pluralism exists in describing PEC efficiency is that several different measures of efficiency (η) exist; each has a different place in PEC research. Four primary measures of efficiency will be discussed here, which can be split into the two main categories:

- Benchmark efficiency (suitable for mainstream reporting of stand-alone water splitting capability)
  - solar-to-hydrogen conversion efficiency (STH)

- Diagnostic efficiencies (to characterize and understand materials system/interface performance)
  - applied bias photon-to-current efficiency (ABPE)
  - external quantum efficiency (EQE) = incident photon-to-current efficiency (IPCE)

Z. Chen et al., *Photoelectrochemical Water Splitting*,
SpringerBriefs in Energy, DOI: 10.1007/978-1-4614-8298-7_2,
© The Author(s) 2013

- internal quantum efficiency (IQE) = absorbed photon-to-current efficiency (APCE).

The definitions described below for PEC devices are in agreement with previous analyses of efficiencies for PEC water splitting found in literature [2]. They are also aligned with those used for photovoltaics (PV) [3] as the two fields share many characteristics in common. Similar to the solar-to-electric conversion efficiency in PV, the only PEC device efficiency that is acceptable for benchmark reporting of stand-alone solar water splitting is STH. The other measures are invaluable scientifically, providing insight into the functionality and limitations of a material or device, but do not represent device STH performance as it relates to PEC targets such as those established by the U.S. Department of Energy, for example in the Fuel Cell Technologies Office Multi-Year Research, Development and Demonstration Plan (http://www1.eere.energy.gov/hydrogenandfuelcells/mypp/index.html). The following sections provide detailed definitions of the different efficiencies, emphasizing their distinctions and proper uses.

2.2 Efficiency Definition for Benchmarking

STH efficiency is the most overarching of all the efficiency metrics as it describes the overall efficiency of a PEC water splitting device exposed to broadband solar irradiance (e.g., Air Mass 1.5 Global illumination [4–6], abbreviated AM 1.5 G) under zero-bias conditions. Zero-bias means that there is no applied voltage between the working electrode (WE) and counter electrode (CE), and all of the energy in the water splitting process is being supplied by sunlight. For direct STH efficiency measurements the WE and CE are operated under short-circuited conditions, so voltages versus the reference electrode or versus the open circuit voltage are not relevant. Additionally, direct STH measurement cannot be made when the WE and CE are compartmentalized and immersed in solutions of different pH. A Nernstian bias of 59 mV per pH unit of difference can arise from the chemical bias between the two solutions [7], and maintaining a pH difference between the two compartments requires the input of chemical energy in addition to the solar energy. Thus for a correct STH efficiency measurement one must have both the WE and CE immersed in the same pH solution (though compartmentalization is still allowable). In addition, the electrolyte should not contain any sacrificial donors or acceptors since the redox (reduction–oxidation) reactions would no longer reflect true water splitting. We do note, however, that designs exist in the literature which may eventually enable sustained operation in systems utilizing chemical inputs in addition to water [8]. These would be subject to different efficiency metrics which take into account the value of all inputs.

For direct solar-to-hydrogen processes, STH efficiency is defined as *chemical energy of the hydrogen produced* divided by *solar energy input from sunlight incident on the process*. The *chemical energy of the hydrogen produced* can be calculated from the rate of hydrogen production (mmol H_2/s) multiplied by the change in Gibbs free energy per mole of H_2 ($\Delta G^0 = 237$ kJ/mol at 25 °C). In some reports, the heat of combustion, or enthalpy, has been used to represent the energy content of H_2 in the STH calculation ($\Delta H^0 = 286$ kJ/mol at 25 °C for the high heating value). However, in standard reporting of STH efficiency, the Gibbs free energy, which accounts for entropy (i.e., $\Delta G = \Delta H - T\Delta S$), should be used as it reflects the maximum amount of electric energy obtainable upon reaction of the hydrogen to form liquid water (e.g., in a fuel cell). Any reporting of STH efficiency should be explicit about the energy basis for the produced hydrogen. In the denominator of the STH definition, the *solar energy input from sunlight* is incident illumination power density (P_{total}, in units of mW/cm^2) multiplied by the illuminated electrode area (cm^2). The illumination source should closely match the shape and intensity of the Air Mass 1.5 Global (AM 1.5 G) G173 standard [9] set forth by the American Society of Testing and Materials (ASTM) [10]. With the above definitions and conditions, the standard definition for STH representative of device operations under one-sun operations can be expressed:

$$\text{STH} = \left[\frac{(\text{mmol } H_2/\text{s}) \times (237{,}000 \text{ J/mol})}{P_{\text{total}} \ (\text{mW/cm}^2) \times \ \text{Area} \ (\text{cm}^2)} \right]_{\text{AM1.5 G}} \tag{2.1}$$

Equation (2.1) calculates the power output (numerator) based on the direct measurement of the true H_2 production rate by an analytical method such as gas chromatography or mass spectrometry; Alternatively, Equation (2.2) uses the relation that power is the product of voltage, current, and the Faradaic efficiency for hydrogen evolution (η_F).

$$\text{STH} = \left[\frac{\left| j_{\text{SC}}(\text{mA/cm}^2) \right| \times (1.23 \text{ V}) \times \eta_{\text{F}}}{P_{\text{total}} \ (\text{mW/cm}^2)} \right]_{\text{AM1.5 G}} \tag{2.2}$$

"Current" is the short-circuit photocurrent density (j_{SC}, mA/cm^2) normalized to the illuminated electrode area and the "voltage" is 1.23 V (ΔE^0 at 25 °C), the thermodynamic water splitting potential (based on ΔG^0). In a PEC device, short-circuit refers to zero voltage in the external circuit, similar to a PV device. However, it is worth noting that majority and minority carriers in a PEC device are equilibrated at separate redox potentials that correspond to the half-reactions for water splitting. It is important to emphasize that the Gibbs free energy is the basis for the numerators in both expressions; The discouraged use of the higher enthalpic heating value of 286 kJ/mol results in a higher voltage of 1.48 V in Eq. (2.2). The denominator is simply the total integrated power input density (P_{total}, in units of mW/cm^2) from the impinging illumination. If input illumination is measured in total power (mW), it must first be normalized to a power density (mW/cm^2) by dividing by the illuminated electrode area.

It is critical to note that Eqs. (2.1) and (2.2) are valid *if and only if* one confirms stoichiometric gas evolution (H_2 and O_2) in the absence of any sacrificial electron donors or acceptors. Equation (2.2) similarly restricts the measured photocurrent to the fraction which corresponds directly to the molar H_2 generation rate; otherwise, Equation (2.2) would give an upper bound efficiency that overestimates the true STH. This assumption is not always valid and laboratories are advised to employ methods for direct identification and quantification of H_2 and O_2 to establish Faradaic efficiencies. Again, if sacrificial electron donors and/or acceptors are used, water is not being split, and the above equations for STH efficiency cannot be applied. Certain diagnostic experiments, for example, utilize organic molecules such as methanol ($E^0_{CO_2, H+/CH_3OH, H_2O} = 0.05$ V) [11] which are much more easily oxidized than water, but these are not indicative of STH water splitting efficiency.

It is worth nothing that the oxidation half-reaction (oxygen evolution reaction in the case of water splitting) must produce H^+ and not necessarily O_2. The H^+ serves as the necessary feedstock for the hydrogen evolution reaction to produce H_2, which is the main desired product from the solar driven reaction. Any number of reagents can be oxidized to provide H^+, such as CH_3OH, HBr, or even H_2 itself. While nearly all of these reagents are easier to oxidize (thereby requiring less total voltage) in comparison to water, they are not sustainable reactants. It is important to produce H^+ from a sustainable and economical source, and water is the best option that fits this need on a global scale.

2.3 Diagnostic Efficiencies

2.3.1 Applied Bias Photon-to-Current Efficiency

Applying a bias between the working and counter electrodes requires a new efficiency value separate from STH since such a value does not reflect a true solar-to-hydrogen conversion process. This merits the definition of an "applied bias photon-to-current efficiency" (ABPE). The application of a bias generally increases the current drawn from the device, but one should be cognizant that applying a bias that exceeds the thermodynamic water splitting potential (1.23 V) brings into question whether or not PEC under these conditions is more advantageous than standard electrolysis in the dark. Given that the ABPE measurement is not a true solar-to-hydrogen measurement, it serves as a diagnostic measurement in materials development.

$$\text{ABPE} = \left[\frac{\left| j_{ph} \left(\text{mA/cm}^2 \right) \right| \times \left(1.23 - |V_b| \right) (V) \times \eta_F}{P_{total} \left(\text{mW/cm}^2 \right)} \right]_{\text{AM1.5 G}} \qquad (2.3)$$

where j_{ph} is the photocurrent density obtained under an applied bias V_b.

As with STH, there are several common pitfalls in ABPE reporting that should be avoided:

1. Reporting a bias only versus a reference electrode and not the counter electrode. Doing so would result in an interface measurement and not a device measurement since a bias versus a reference electrode excludes the second half-reaction occurring at the counter electrode.
2. Using a sacrificial donor or acceptor in the electrolyte. While this may be extremely useful in diagnosing interfacial charge transfer limitations, it fails to represent a stand-alone water splitting process.
3. Using a chemical bias (e.g., a 2-compartment cell with electrolytes at different pH).

2.3.2 Incident Photon-to-Current Efficiency (IPCE)/External Quantum Efficiency (EQE)

The Incident Photon-to-Current Efficiency (IPCE) is one of the most important diagnostic figures of merit for PEC devices; it describes the photocurrent collected per incident photon flux as a function of illumination wavelength. Ideally, the researcher can integrate the IPCE data over the solar spectrum in order to *estimate* the maximum possible STH efficiency for that device, but only for the IPCE data collected under zero-bias (2-electrode, short-circuit) conditions. IPCE under an applied bias is not considered a valid estimate for STH, but it is still a very useful diagnostic tool which gives insight into the PEC material properties. IPCE takes into account efficiencies for three fundamental processes involved in PEC, as illustrated in Fig. 2.2 in the "Introduction" chapter: photon absorptance, defined as the fraction of electron–hole (e^-/h^+) pairs generated per incident photon flux (η_{e^-/h^+}), charge transport to the solid–liquid interface $(\eta_{transport})$, and the efficiency of interfacial charge transfer $(\eta_{interface})$. This assumes that the counter electrode is not limiting current flowing through the circuit.

$$\text{IPCE} = \text{EQE} = \eta_{e^-/h^+}\,\eta_{transport}\,\eta_{interface} \tag{2.4}$$

Strictly speaking, IPCE is identical to EQE. In PV devices $\eta_{interface}$ is often close to or equal to 1 as charges are extracted to a metal that forms an ohmic contact with the semiconductor (however, avoiding interface recombination also plays a significant role in the optimization of PV devices). In PEC, however, interfacial charge transfer kinetics for the water splitting reaction (hydrogen and oxygen evolution) are often sluggish; thus the probability of electron or hole transfer across the solid–liquid interface $(\eta_{interface})$ is generally not unity.

In a PEC system, IPCE is usually obtained from a chronoamperometry (potentiostatic) measurement. In this system, a bias can be applied between the sample/working electrode versus a counter electrode (2-electrode experiment) or a reference electrode (3-electrode experiment) while measuring the current that arises from subjecting the PEC electrode to monochromatic light at various wavelengths. The difference between the steady state current under monochromatic illumination and the steady state background current is the photocurrent that arises due to redox reactions occurring at the surface of the working electrode. The IPCE corresponds to the ratio of this photocurrent (converted to an electron rate) versus the rate of incident photons (converted from the calibrated power of a light source). Calibrated monochromated light (alone, or superimposed over a background illumination level in the case of a white light bias experiment) should be used for this experiment to give IPCE as a function of wavelength λ (nm).

$$\text{IPCE}(\lambda) = \text{EQE}(\lambda) = \frac{\text{electrons/cm}^2/\text{s}}{\text{photons/cm}^2/\text{s}} = \frac{\left| j_{\text{ph}}(\text{mA/cm}^2) \right| \times 1239.8(\text{V} \times \text{nm})}{P_{\text{mono}}(\text{mW/cm}^2) \times \lambda(\text{nm})}$$

$$(2.5)$$

where 1239.8 V $\times$ nm represents a multiplication of h (Planck's constant) and c (the speed of light), P_{mono} is the calibrated and monochromated illumination power intensity in mW/cm^2, and λ (nm) is the wavelength at which this illumination power is measured.

IPCE is an extremely useful number as it yields device efficiency in terms of "electrons out per photons in" (as opposed to "power out per power in"). This takes into account the spectral variation of incident photons at each energy. For example, if an equivalent number of 400 nm photons and 600 nm photons impinge on a sample and an equal number of electrons is collected at each wavelength, the IPCE for both wavelengths is identical, despite the higher energy of the 400 nm photons. In the context of PEC water splitting, IPCE describes the maximum possible efficiency with which incoming radiation can produce hydrogen from water, with an implicit assumption that all electrons are used for the evolution of H_2 (and holes for evolution of O_2) instead of other byproducts or corrosion (i.e. Faradaic efficiencies for H_2 and O_2 evolution should be ~ 100 %). To re-emphasize the point, this assumption is not always valid and laboratories are advised to employ methods for direct identification and quantification of H_2 and O_2 to establish the Faradaic efficiency for water splitting. In the absence of a known Faradaic efficiency, it is important to keep in mind that the measured performance may be artificially inflated.

To avoid confusion between IPCE and solar-to-hydrogen conversion efficiency (STH), we should note three critical differences. First, in contrast to IPCE (electrons out/photons in), STH describes efficiency in terms of power (power out/power in). Secondly, whereas IPCE measurements can be conducted with any calibrated and monochromated illumination source (i.e. it need not be AM 1.5 G illumination that is monochromated as long as the number of impinging photons at each wavelength are counted), STH requires the use of broadband solar-simulated illumination. The reader is referred elsewhere [5] for a discussion of solar simulation using laboratory illumination sources. The integration of IPCE over the entire solar spectrum can provide an estimation of the maximum possible STH, if (and only if) no applied bias is used in the IPCE measurement. Lastly, conducting IPCE experiments with an applied bias is allowable whereas STH requires true zero-bias conditions. Of course, the authors of a publication must be explicit as to the bias applied in an IPCE experiment. The bias versus the counter electrode is the most pertinent, but a bias versus the reference electrode may alternatively be used (note that 0 V vs. a reference electrode is not a zero-bias condition).

2.3.3 Absorbed Photon-to-Current Efficiency (APCE)/ Internal Quantum Efficiency (IQE)

PEC device efficiencies as measured by IPCE/EQE or STH implicitly include losses from impinging photons that are reflected or transmitted. To understand the inherent performance of a material, it is often helpful to subtract these losses and measure efficiency based only on photons *absorbed*. This is known as the absorbed photon-to-current efficiency (APCE), which describes the photocurrent collected per incident photon absorbed. APCE is synonymous with internal quantum efficiency (IQE). This is a particularly useful quantity to measure when studying thin films, since it helps to determine the optimum balance between maximal pathlength for photon absorption versus minimal effective e^-/h^+ transport distance within the material.

$$\text{APCE} = \text{IQE} = \frac{\text{IPCE}}{\eta_{e^-/h^+}} = \eta_{\text{transport}}\eta_{\text{interface}} \tag{2.6}$$

where η_{e^-/h^+} is absorptance, defined as the fraction of electron–hole (e^-/h^+) pairs generated per incident photon flux. Absorptance is estimated from Beer's Law, which defines the absorbance (A) of a sample as the logarithmic ratio of the measured output light intensity (I) versus the initial input light intensity (I_0). This value is readily determined experimentally by UV–Vis Spectroscopy, which provides a measurement of the number of photons absorbed per incident photon impinging on the sample. The assumption is that the number of e^-/h^+ pairs generated equals the number of photons absorbed.

$$A = -\log\left(\frac{I}{I_0}\right) \tag{2.7}$$

$$\eta_{e^-/h^+} = \frac{I_0 - I}{I_0} = 1 - \frac{I}{I_0} = 1 - 10^{-A} \tag{2.8}$$

η_{e^-/h^+} describes the first of the three fundamental processes in PEC and establishes the maximum attainable PEC efficiency for the device, since $\eta_{\text{transport}}$ and $\eta_{\text{interface}}$ can never be greater than unity. If a material exhibits a low IPCE but a high η_{e^-/h^+}, then the reduction in efficiency arises from either poor transport or poor interfacial kinetics (or both). One way to decouple the two limiting factors is to re-run the IPCE experiment with a more facile redox couple for interfacial charge transfer than the hydrogen evolution reaction (HER) and the oxygen evolution reaction (OER). This will set $\eta_{\text{interface}} \approx 1$ and ensure that any measured difference between η_{e^-/h^+} and IPCE (with the more facile redox couple) arises solely from $\eta_{transport}$. Ideally, $\eta_{transport}$ should not have a strong dependence on the redox couple chosen. However, $\eta_{interface}$ will be dictated by this decision and thus an appropriate redox couple must be judiciously selected such that the probability of interfacial charge transfer to electron acceptors and from electron donors in solution is effectively 100 % for photo-excited charges that reach each electrode surface.

Combining the equations for determining IPCE and η_{e^-/h^+} experimentally, APCE can be derived as follows:

$$\text{APCE}(\lambda) = \text{IQE}(\lambda) = \frac{\left|j_{ph}(\text{mA/cm}^2)\right| \times 1239.8(V \times \text{nm})}{P_{mono}(\text{mW/cm}^2) \times \lambda(\text{nm}) \times (1 - 10^{-A})} \tag{2.9}$$

2.4 Half-cell Metrics

Though a true device efficiency can never be obtained without fully accounting for both the working and counter electrodes (i.e. in a 2-electrode configuration), half-cell measurements of a working electrode vs. a reference electrode (3-electrode configuration) are nonetheless the most prevalent form of experimentation utilized by PEC researchers. Such measurements yield important insight into material and interface properties, and reflect the simple fact that researchers do not assemble a full device without first understanding its individual components. The following discussion highlights the metrics that allow a researcher to assess the viability of a photoactive material for integration into a full PEC device. However, it bears repeating:

No single half-cell (3-electrode) measurement can produce a valid device efficiency

The most common example is the use of what appears to be the ABPE with an invalid voltage as per the following equation.

$$\text{Half-cell"Efficiency"} = \left[\frac{j_{ph}(\text{mA/cm}^2) \times \left|E_{ref} - E\right|(V)}{P_{total}(\text{mW/cm}^2)}\right]_{AM1.5G} \tag{2.10}$$

where E_{ref} is some reference potential. In some cases, the voltage used is referenced to the reversible potential for the reaction of interest such as HER/OER. In other cases, researchers replace $(E_{ref} - E)$ with a voltage bias vs. an arbitrary reference electrode (e.g., 0.3 V vs. Ag/AgCl). It is important to keep in mind that neither approach is correct since these are not full cell voltages. There are many reports of such half-cell efficiency numbers in the PEC literature that do not represent a true closed cycle reaction pathway (e.g., water splitting). This measure of "efficiency" inflates the apparent performance of a material, misleads other researchers, and has no greater practical value since it cannot be converted into a meaningful device performance metric. However, certain components of the equation do yield useful measures of performance, as will be described in Chapter "Flat-Band Potential Techniques".

2.5 Summary of Efficiency Definitions

There are many efficiency values that can be reported for PEC. We have described the four most important to consider in PEC research: STH, ABPE, IPCE/EQE, and APCE/IQE. It cannot be emphasized enough that STH efficiency is the most important of all efficiency measurements and great care must be taken to ensure that this measurement is done correctly and with accurate understanding of the photoelectrochemistry it reflects. STH is the only efficiency that can be used to determine water splitting H_2 production efficiency and that should be used as a benchmark value to compare different PEC candidate materials. The definition of STH was presented along with the strict conditions on how measurements and calculations should be conducted. The other efficiencies, namely IPCE, APCE, and ABPE, are valuable for understanding and improving material performance, i.e. diagnostic measurements. However, high values obtained for IPCE, APCE, and ABPE do not necessarily translate to high values of STH efficiency, the true PEC efficiency that serves as a benchmark for materials.

References

1. O. Khaselev, A. Bansal, J.A. Turner, High-efficiency integrated multijunction photovoltaic/ electrolysis systems for hydrogen production. Int. J. Hydrogen Energy **26**, 127–132 (2001)
2. O.K. Varghese, C.A. Grimes, Appropriate strategies for determining the photoconversion efficiency of water photo electrolysis cells: a review with examples using titania nanotube array photoanodes. Sol. Energy Mater. Sol. Cells **92**, 374–384 (2008)
3. Quantum Efficiency Measurements, U.S. Department of Energy (2005), http://www1.eere. energy.gov/solar/quantum_efficiency.html. Accessed 1 Jan 2010
4. T. Bak, J. Nowotny, M. Rekas, C.C. Sorrell, Photo-electrochemical hydrogen generation from water using solar energy. Materials-related aspects. Int. J. Hydrogen Energy **27**, 991–1022 (2002)
5. H. Mullejans, A. Ioannides, R. Kenny, W. Zaaiman, H.A. Ossenbrink, E.D. Dunlop, Spectral mismatch in calibration of photovoltaic reference devices by global sunlight method. Meas. Sci. Technol. **16**, 1250–1254 (2005)

6. G.P. Smestad, F.C. Krebs, C.M. Lampert, C.G. Granqvist, K.L. Chopra, X. Mathew, H. Takakura, Reporting solar cell efficiencies in solar energy materials and solar cells. Sol. Energy Mater. Sol. Cells **92**, 371–373 (2008)
7. A.J. Nozik, Photoelectrolysis of water using semiconducting TiO_2 crystals. Nature **257**, 383–386 (1975)
8. M. Unlu, J. Zhou, P.A. Kohl, Hybrid anion and proton exchange membrane fuel cells. J. Phys. Chem. C **113**, 11416–11423 (2009)
9. A.B. Murphy, P.R.F. Barnes, L.K. Randeniya, I.C. Plumb, I.E. Grey, M.D. Horne, J.A. Glasscock, Efficiency of solar water splitting using semiconductor electrodes. Int. J. Hydrogen Energy **31**, 1999–2017 (2006)
10. ASTM Standard G173, 2003e1, *Standard Tables for Reference Solar Spectral Irradiances: Direct Normal and Hemispherical on 37° Tilted Surface*, ASTM International, West Coshohocken (2003). doi:10.1520/G0173-03E01, www.astm.org
11. M.A.A. Schoonen, Y. Xu, D.R. Strongin, An introduction to geocatalysis. J. Geochem. Explor. **62**, 201–215 (1998)

Chapter 3
Experimental Considerations

Standardized characterization of PEC materials and photoelectrodes requires careful attention to experimental methods in sample preparation and testing setups. Fundamental experimental considerations are discussed in this chapter.

3.1 Electrode Preparation

3.1.1 Electrode Preparation Considerations

Researchers must take care when preparing electrodes for PEC testing to ensure that they only measure the performance of the material of interest without contributions from the surrounding system which can arise from poor electrode preparation. For example, proper sealing is critical in order to keep the electrolyte from contacting anything but the intended surface of the electrode. It is also important to minimize electrical resistance and ensure a good ohmic contact between the semiconductor and the substrate. Other factors to be considered in sample preparation include the method of deposition, the substrate type, as well as the crystallinity, and thickness of the material. These factors, along with some examples of electrode mounting, are discussed below.

3.1.2 Photoactive Semiconductor Material

3.1.2.1 Deposition Techniques

Many deposition methods are available for fabricating semiconductor electrodes. However, each technique has advantages and drawbacks that should be kept in mind when choosing a fabrication method. For instance, monocrystalline materials with high PEC performance can often be obtained via molecular epitaxy processes.

Z. Chen et al., *Photoelectrochemical Water Splitting*,
SpringerBriefs in Energy, DOI: 10.1007/978-1-4614-8298-7_3,
© The Author(s) 2013

However, such techniques require highly specialized substrates and/or large thermal budgets that might not be compatible with the more common, low cost metallic foils, or glass substrates. Atomic layer deposition processes can provide precise control over material thickness; however, such control requires low deposition rates, which would reduce device throughputs when thicker films are desired. Thus, a balance between semiconductor PEC performance and the cost, speed, and scalability of the material fabrication method must be considered. Commonly used techniques to fabricate semiconductor photoelectrodes include:

- *Physical vapor deposition (PVD).* This class of deposition refers to evaporation and sputtering processes performed in vacuum chambers. The material of interest comes from a solid source that is either heated (evaporation) or etched using plasma (sputtering) yielding a vapor that condenses on the substrate to form the electrode. Multiple sources can be used at once to form alloys. PVD processes usually lead to polycrystalline films.
- *Molecular beam epitaxy (MBE).* This technique is another class of PVD processes. The main difference from the traditional PVD process resides in the control of the vapor flow, which, for MBE allows the growth of monocrystalline films (*epi* "above" and *taxis* "in ordered manner"). To achieve such a highly crystalline film, a crystalline substrate is required, the substrate must be heated to several hundred degree Celsius during deposition, and the precursor vapor must travel through a very high vacuum (10^{-8} Pa) at relatively low flow.
- *Chemical vapor deposition (CVD).* In CVD, the material's components come from the decomposition of one or more volatile chemical precursors that decompose and/or react on the substrate. Depending on deposition conditions (temperature and pressure) and the precursor nature, different terminologies are used to define CVD processes. For example, MOCVD is metal organic CVD, LPCVD is low-pressure CVD, ALCVD is atomic layer CVD, HWCVD is hot-wire CVD, and PECVD is plasma-enhanced CVD. These processes can lead to single crystal, amorphous, or polycrystalline films.
- *Electrodeposition.* In this process, a conductive substrate is placed in an electrolyte solution (typically aqueous) that contains a salt of the material of interest. When an electrical potential is applied between the substrate and a counter electrode, redox chemistry takes place at the surface of the substrate which deposits material. Complex pulse trains and/or high-pulse frequencies are sometimes used to direct current flow and favor desired reactions. A postsynthesis calcination is often performed to reach a desired material phase. Electrodeposition is restricted to deposition of electrically conductive materials and produces polycrystalline and amorphous films. This process is also appropriate for thin film surface treatment of PEC electrodes, such as electrocatalyst deposition.

This list is by no means exclusive. Other available deposition techniques include sol–gel, powder pressing, etc.

3.1.2.2 Crystallinity

Crystallographic properties play an important role in thin film electronic properties such as the material's band gap energy, the Fermi level position, and the carrier mobilities. As described above, the choice of deposition process generally determines the crystallographic properties (amorphous, polycrystalline, or monocrystalline) of the thin film. To achieve highly crystalline semiconductor materials (such as $GaInP_2$), MOCVD or MBE are generally preferred [1]. These deposition techniques typically lead to higher performance materials, but are more costly to perform. Therefore, it is important to select the right deposition method to achieve a balance between high performance and low fabrication cost in PEC films. For instance, materials such as amorphous-SiC can be deposited using low-cost deposition processes such as PECVD [2]. Metal-oxide materials (Fe_2O_3, TiO_2, and WO_3) can be deposited using moderately priced PVD processes (sputtering, evaporation) or low-cost electrochemical depositions [3–5].

3.1.2.3 Material Thickness

To maximize charge carrier generation and collection, the thickness of the semiconductor should be on the order of the optical penetration depth (α^{-1}, where α is the absorption coefficient). Strictly, a thickness of α^{-1} corresponds to 63 % absorption of incident light. Excess material may add ohmic resistance to the cell and/or enhance the likelihood of carrier recombination, with little gain in absorption. For a direct band gap material with $\alpha \sim 10^6 \, cm^{-1}$ in the visible (~ 600 nm), a 1-μm thick film is generally sufficient. However, for an indirect band gap material with a lower $\alpha \sim 10^4 \, cm^{-1}$, a thickness of a few hundred micrometers may be required. Note that materials with poor majority or minority carrier transport may need to be thinner than their optical penetration depth for optimal performance. Very thin micro/nanostructured films may have pore channels that expose the conductive substrate to the electrolyte and lead to shunting. To mitigate shunting, a very thin ($\sim$ few nm) tunneling barrier layer of oxide may be deposited on the conductive substrate prior to film deposition in some cases.

3.1.3 Substrate Considerations

3.1.3.1 Ohmic Contact

To ensure optimal charge carrier transfer at the interface between the semiconductor and the substrate, it is essential to form a low-loss electronic contact (e.g., an ohmic or tunnel junction contact). An ohmic contact, as the name implies, follows Ohm's law that current is a linear function of applied potential. Formally, ohmic contacts do not rectify an electric current and do not inject minority carriers into the bulk of the semiconductor, but can have a moderate resistance. Generally, in order for the contact to be ohmic, the substrate must enrich the majority carriers at the interface to a level greater than in the semiconductor bulk. For a p-type semiconductor, a conductive material with a work function (ϕ_M) larger than the semiconductor work function (ϕ_S) is typically required. Gold ($\phi_M = 5.3$ eV) [6] is a candidate for a large work function ohmic contact. In this configuration, holes gather at the surface forming an accumulation layer and the semiconductor behaves as a metal at the junction. Similar arguments apply to n-type materials where a conductive material with a smaller work function than that of the semiconductor is required. Aluminum ($\phi_M = 4.3$ eV) [7] is a small work function candidate. It is worth mentioning that the work function of a given material depends on several factors including chemical composition [7] and crystallographic orientation [8]. Thus, the selection of a contact material can be guided by reported work function values but is best done through a careful characterization of the electronic properties of the contact/semiconductor interface [9]. In particular, researchers should verify the formation of an ohmic contact at the semiconductor/metal interface and not a Schottky barrier which reflects majority carriers back into the bulk of the semiconductor and impedes charge transport. Additionally, some substrates may alloy with the semiconductor, leading to a different material with its own work function. This alters the contact, for better or worse, which needs to be verified experimentally.

3.1.3.2 Compatibility with Deposition Processes and Integration Scheme

Materials such as stainless steel, aluminum, titanium, and molybdenum can form good ohmic contacts with semiconductor films, as long as the conditions described in the previous section are satisfied. However, in some circumstances, those materials are not compatible with deposition processes. For epitaxial syntheses such as MBE, MOCVD, or metal–organic vapor phase epitaxy (MOVPE), single crystal semiconductor substrates are chosen so that their lattice constants are close to those of the desired epilayer composition. GaP, GaAs, and Si are generally used in this case, and they are degenerately doped to support current flow with minimal resistance. Materials such as sapphire or SiC can be used when a high temperature process is required. When low temperature processes (<600 °C) are employed, glass substrates coated with fluorine-doped tin oxide (FTO) or zinc oxide substrates can be used. These significantly conductive substrates are interesting when PEC integration schemes require transparent conductive films. Thus, either front or back-side illumination can be employed to characterize the material. A list of substrates compatible with select semiconductors is presented in Table 3.1.

Table 3.1   List of substrates compatible with some semiconductors

Semiconductor	Deposition process (Temperature)	Known compatible substrates
CGSe	Evaporation (550 °C)	FTO [10]
WO_3	Sputtering (300 °C)	FTO [10], ITO [11]
$GaInP_2$	MOCVD	GaAs [12]
GaPN	MOCVD	GaP [13], Si [13]
InGaN, $GaInP_2$	MBE	SiC, sapphire
Fe_2O_3	Electrodeposition + calcination (450–700 °C)	FTO [14], glass/Ti/Pt [5]
	Sputtering (450 °C)	
Cu_2O	Electrodeposition	Ti [15], ITO [16], FTO
a-SiC	PECVD	ZnO_2 [2], FTO [2], Stainless steel [2], Cr [2]

3.1.3.3 Connecting the Substrate to the Electrical Circuit

The method employed to connect the semiconductor to the electrical circuit depends first and foremost on the experimental test setup. When a conventional laboratory cell is used (e.g., open beaker or three-port glass cell), a wire must be attached to the conductive substrate upon which the semiconductor is deposited. Such connection can be obtained by soldering wires (copper or similar electrical leads) onto the conductive substrate, using silver paint or depositing indium. If the substrate is made of a fully conducting material, such as stainless steel or doped silicon, the connection can be made directly to the back side of the sample. For glass coated FTO substrates, the nonconductive glass necessitates that either some portion of the FTO be left uncovered during the semiconductor deposition, or a portion of the semiconductor must be etched away before testing. Finally, a nonconductive epoxy sealant may be applied to prevent exposure of the electrolyte to nonsemiconductor components (such as the substrate, wiring, or solder). The epoxy also provides mechanical support to prevent the electric wire from accidental detachment. It is worth mentioning that the epoxy should be compatible with the electrolyte over the range of pH used to test the PEC electrode (i.e., does not appreciably corrode or leach electrochemically active compounds into solution for the duration of the test). The epoxy should also be opaque to provide control over the illuminated area.

3.1.4 PEC Electrode Connections

Several methods are commonly used to prepare PEC electrodes, but many similar techniques should work as long as they fulfill some basic requirements: (1) a large area ohmic contact to provide uniform potential distribution, (2) electrical and chemical isolation of all conductive hardware (wires, contacts, electrode edges, etc.) from the electrolyte except for the planar working electrode (WE) surface. In this section, two different electrode mounting techniques are presented as examples.

3.1.4.1 Electrode Connection Using Insulated Electrical Wires

Connection to an electrode using insulated electrical wires is performed in five steps:

(1) Depending on electrode dimensions and conductivity of the substrate, one or two wires may be required to achieve a uniform potential distribution and enhance charge collection (Fig. 3.1a).

(2) Both ends of each wire are stripped and one end is soldered onto the outer edge of the uncoated conductive part of the substrate. Indium solder is often a good choice for soldering the lead wires to the substrate due to its high conductivity and good adhesion to surfaces. For better adhesion, a thin layer of indium can be deposited first onto the substrate (Fig. 3.1b). Next, more indium is used to attach the wires onto the substrate (Fig. 3.1c). Alternatively, silver paint can be used, in which case the sample should dry in air for a few hours or annealed at 80 °C for 20–30 min.

(3) The indium, any uncoated substrate and uninsulated parts of the wires must be covered with epoxy resin (e.g., Resinlab EP1290 or HYSOL 9462). A sufficient thickness of epoxy should be used to minimize possibility of pinhole formation.

(4) All epoxy-covered electrodes should cure for at least 2 h at room temperature, although curing time is dependent on the type of epoxy. Electrodes can then be placed in an oven for 2 h at 55–80 °C to accelerate the curing process.

(5) As epoxy can flow during the hardening process, a touch-up may be required to re-cover any conductive surfaces that may have been exposed (Fig. 3.1d). If aggressive electrolytes are required, the annealed epoxy must be covered with a thin layer of HYSOL E-120 HP epoxy, mixed 90 min prior to use, and left to rest in air for at least 3 h.

Note. Many epoxies contract as they cure. This contraction may lead to cracking of the semiconductor film, exposing the underlying metallic contact to the electrolyte. In this case, an O-ring seal to the electrolyte is recommended.

Another variation of making an electrode connection is shown in Fig. 3.2. In this case, bare copper wire is inserted into a glass tube and coiled at the end. The coiled wire is attached to the back side of the sample using silver paint and allowed to dry at 80 °C for 20–30 min. The entire assembly is then encased in epoxy and annealed at 80 °C for 2 h.

3.1.4.2 Electrode Connection Using Copper Tape and O-Ring

This method is very useful when multiple samples are arranged in a library array, with the test cell moving from one sample to another to perform PEC electrode characterizations, or when epoxy encasement is undesirable. To confine the electrolyte over the semiconductor area of interest, an O-ring is used to seal the contact between the electrode and test cell that contains the electrolyte. The O-ring should

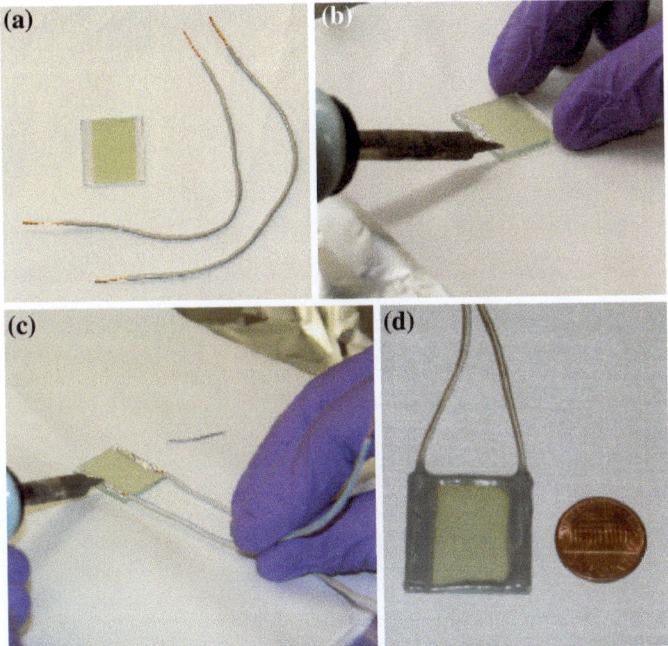

Fig. 3.1 Main steps of PEC electrode connection using plastic insulated wires. (**a**) WO_3 film on FTO and wiring. (**b**) Adding a coat of indium solder to the uncovered FTO. (**c**) Soldering the wires to the indium. (**d**) Coating the electrically conductive wiring and substrate with nonconductive epoxy

be made of materials (such as Viton or Teflon-coated silicone, the former is incompatible with organic solvents) that are durable in the testing condition. Minimal leaching of organic or other contaminants is desired to prevent parasitic reactions that compete with O_2 and H_2 generation and/or contamination of the H_2/O_2 reaction sites and lower the efficiency. Generally, hard O-rings are not used because they are difficult to seal and may damage the sample. The presence of an O-ring ensures that the edges of the photoelectrode do not come in contact with the solution, so epoxy encasement is not necessary. It is worth mentioning that this contact method allows further characterizations to be performed after electro chemical testing. Postelectrochemical analysis such as ultra-high vacuum (UHV) surface characterization may not be feasible with the epoxy present. Copper tape with an electrically conductive adhesive (e.g., McMaster-Carr 76555A712) can be used to make an electrical connection to the conductive substrate of the photo-electrode. Then, a protective layer of Teflon Tape (e.g., McMaster Carr 76475A31) is applied on top of the copper tape. This decreases the chance of sample contamination by Cu ions in the event of electrolyte leakage onto the copper tape.

Figure 3.3 shows two samples mounted using this method. The copper tape is attached to an uncovered area of the conductive substrate (no semiconductor

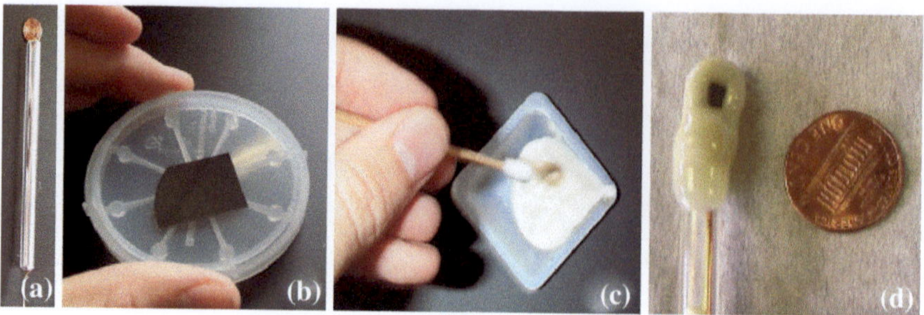

Fig. 3.2 Connecting an electrode with Cu wire. Bare Cu wire is inserted into a glass tube and coiled at the end to increase surface area (**a**). The back side of the sample (**b**) is connected to the Cu coil using silver paint, followed by the addition of HYSOL 9462 epoxy (**c**) to encase the entire assembly. The finished product is shown in (**d**)

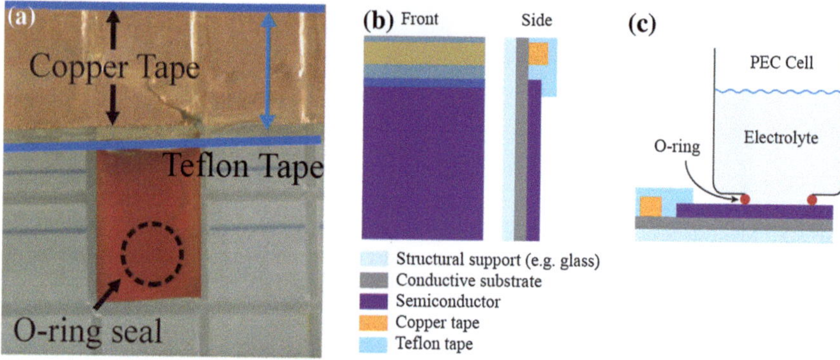

Fig. 3.3 (**a**) Electrode preparation using copper tape contact and protective Teflon tape. O-ring contact position indicated by *dashed circle*. (**b**) Plan-view schematic of the electrodes; the top portion of the conductive substrate is left uncoated for contact with Cu tape, which is then covered with Teflon tape. (**c**) Side view of the electrode in a PEC cell with an O-ring seal

applied to this area). The Teflon tape is applied over the copper tape (note the Teflon tape is wider than the copper tape), and also contacts the sample ($\sim$ 1–2 mm) to form a good seal and to prevent Cu dissolution in the event of electrolyte leakage. A dashed circle is drawn to show where the O-ring is seated on the sample for PEC testing.

3.1.5 Electrode Surface Area Determination

PEC electrode performance must be normalized to the planar projected electrode surface area, which makes accurate determination of the electrode surface area very important. Several techniques can be used to determine the planar illuminated area. Two common approaches are presented here.

Fig. 3.4 Digital picture of three PEC electrodes. The ruler is used to set the scale for surface area determination

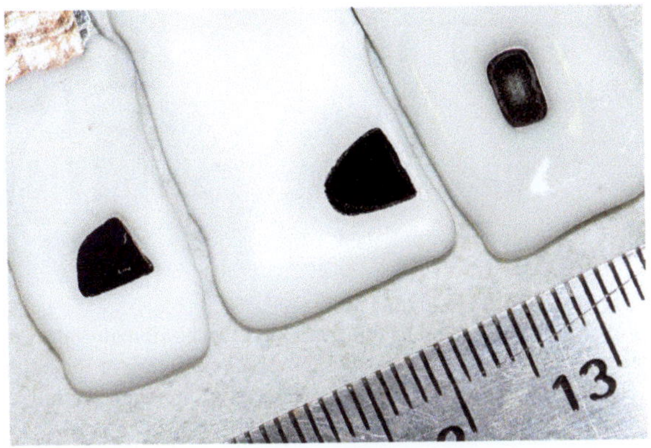

3.1.5.1 Digital Picture

Surface area can be determined by counting the number of pixels on a digital picture of an electrode. The picture can be either obtained using a scanner ($\geq$300 dpi is preferable), photocopier, or a digital camera. A ruler must appear on the picture to set the scale, as shown in Fig. 3.4. Pictures can then be analyzed using image-processing software, such as "Image SXM," a freeware package available for Macintosh computers. A free Windows®-based program is "Image J." The PEC cell active area exposed to solution is defined manually using a pointer. Surface area determination with this software works best with clearly defined regions of contrast between the electrode and its boundary, such as the black and white interface seen in Fig. 3.4. Alternatively, the photocopy and weighing method can be used to determine the geometric surface area of the electrode.

3.1.5.2 Delimitation of the Geometric Surface Area by an O-Ring

By using an O-ring seal between the sample and the main testing chamber in a compression cell setup, the exposed planar geometric area of the sample and the illumination spot are defined by the O-ring. One should use as large of an O-ring as possible for a given sample size to minimize errors in area determination from electrolyte seepage underneath the O-ring beyond the inner diameter or from shrinkage of the inner diameter during compression. Digital pictures are not required to determine the surface area and inter-sample data comparison is simple and quantitative. Figure 3.5c shows the bottom of a combinatorial synthesis and screening instrument (the "combi-probe") [17]. Here, the O-ring is pressed against the sample using bolts on the compression cell to create a sealed chamber.

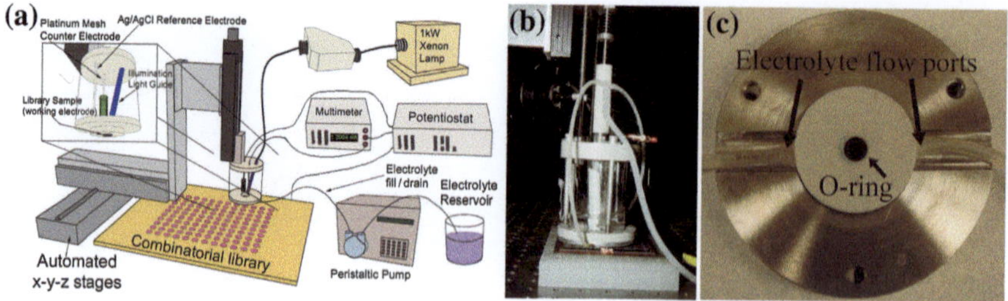

Fig. 3.5 (**a**) Schematic diagram of PEC combinatorial system. (**b**) Photograph of the three-electrode cell which serves as the probe. (**c**) Picture of the bottom of the combinatorial probe cell. The inlet/outlet is used to pump electrolyte to/from the cell as it moves from sample to sample

3.2 Cell Setup and Connections for Three- and Two-Electrode Configurations

3.2.1 Basic Photoelectrochemical Test Setup

In this section, we will explore the basic setup for PEC testing. A general PEC cell contains a working electrode (WE) and a counter electrode (CE) (forming a two-electrode configuration) with an optional reference electrode (RE) (in a three-electrode configuration). A schematic of a general laboratory PEC cell is shown in Fig. 3.6 for both a single compartment setup (accommodating all three electrodes)

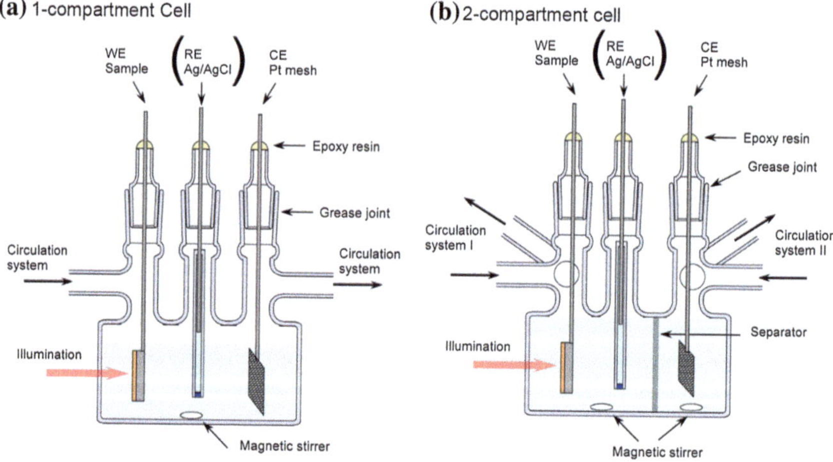

Fig. 3.6 PEC cell in a single (**a**) and double (**b**) compartment configuration [20]. Indicated are ports for the working electrode (WE), counter electrode (CE), and for inlet circulation for gas detection. An optional port for a reference electrode (RE) is shown for three-electrode experiments. Stirring is also optional

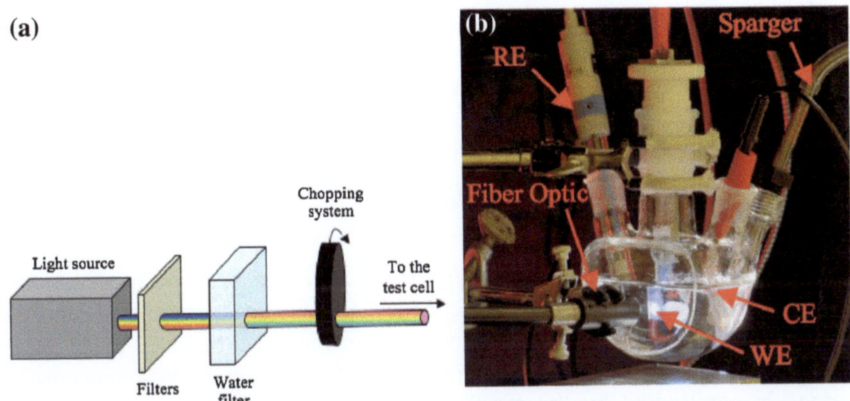

Fig. 3.7 (**a**) Sketch of a horizontal photoelectrochemical test setup and (**b**) picture of test cell including the RE (*left*), the WE illuminated by the incident light (*middle*), and the CE (*right*)

and double compartment setup (with the WE explicitly separated from the CE). In both cases, it is also possible to place the RE into an additional separate compartment connected using a salt bridge (not shown). Many variations on the basic setup are possible and can be found in the literature [18]. The impact of cell design on overall device performance has also been discussed in-depth, providing important guidance to PEC researchers [19].

The basic PEC test setup consists of a light source, light filters, a chopper or shutter (optional), a photoelectrochemical cell, and an electrical measurement tool. The diagram presented in Fig. 3.7a represents a PEC test setup mounted horizontally. Several types of light bulbs are available, but tungsten or xenon bulbs are usually preferred for their broad range of energies, including the ultra-violet spectral region. When AM 1.5 G illumination is required, a series of infrared, neutral density, or cutoff filters might be required to achieve a more representative spectral distribution. Ideally, the resulting light spectrum should be characterized on a regular basis with a spectroradiometer or other calibration tool. If intermittent light is necessary to perform chopped or frequency-dependent characterizations, a chopping system (controlled shutter or rotating wheel) can be employed. To lower the intensity of the infrared portion of the light source, a water filter can be placed in the beam path (this will also minimize heating of the cell).

The three- or two-port photoelectrochemical cells (e.g., Fig. 3.7b) should be made of material that is transparent to the illumination spectrum of interest. If UV irradiation is required, quartz windows must be used in the cell, as opposed to borosilicate glass which generally absorbs wavelengths below ~ 360 nm. The distances between all electrodes should be minimized to limit the effect of electrolyte resistance on the electrochemical test. For electrical measurements, a low impedance ammeter can be used to measure the short-circuit current density.

However, more sophisticated measurement tools such as potentiostats or source meters are required to measure the photocurrent density as a function of applied potential (particularly in measurements referenced to the RE).

3.2.2 Selecting the Counter Electrode

The choice of the CE is based on the type of WE. The size, or more specifically the electrochemically active surface area, of the CE plays an important role in PEC testing, especially in two-electrode configurations, and can be made much larger than the WE. An oversized CE will ensure that the reaction at the semiconductor surface is not limited by the reaction kinetics at the CE surface. For n-type semiconductors, H^+ is reduced into H_2 via the hydrogen evolution reaction (HER) at CE. In this case, large surface area Pt foils (e.g., Pt black) or meshes can make good CEs. For a p-type WE, H_2O is oxidized into H^+ and O_2 via the oxygen evolution reaction (OER) of the CE. In this case, Pt is not suitable since it is a poor OER catalyst. Oxides such as RuO_2 (or hydrates: $RuO_2 \bullet xH_2O$) are better candidates and can reduce overpotentials compared to the noble metals. Other materials such as IrO_2 could also serve as counter electrodes. It is important that Pt CEs are not utilized in studies of p-type WEs (and RuO_2 CEs in n-type WEs) as the catalyst may slowly dissolve and deposit onto the WE, inflating the performance. As a final note, development of nonprecious metal catalysts for HER and OER is highly desired for cost-effective PEC water splitting, since Pt, RuO_2, and IrO_2 are expensive.

3.2.3 Selecting the Reference Electrode

3.2.3.1 Reference Electrodes for Acidic Solution

REs are required in PEC testing in order to measure the potential of the WE on a well-defined electrochemical scale, typically versus the normal hydrogen electrode (NHE). Several REs are available, but saturated calomel electrodes (SCE, $E^0_{SCE} = +0.241$ V vs. NHE at 25 °C) and silver–silver chloride in saturated KCl (Ag/AgCl, $E^0_{Ag/AgCl,\ sat'd\ KCl} = +0.197$ V vs. NHE at 25 °C) electrodes are easiest to use and usually preferred for tests in acidic solutions for the sake of convenience [21]. Note that Ag/AgCl reference electrodes with saturated KCl are not compatible with perchloric acid ($HClO_4$) electrolytes since $KClO_4$ will precipitate in the frit. Replacement with NaCl solution may be a suitable solution.

3.2.3.2 Reference Electrode for Basic Solution

The SCE and Ag/AgCl reference electrodes can generally be used in a wide range of pH values. However, immersion for a long period of time in basic solutions could damage the electrode frit. Thus, Hg/HgO reference electrodes ($E^0_{\text{Hg/HgO, KOH 20\%}} = +0.097$ V vs. NHE at 25 °C) are preferred when basic solutions are required. These REs are often made out of plastic rather than glass, which can dissolve in basic solutions.

3.2.3.3 Measurement Reproducibility

Since several different REs may be used in a lab and their potential may drift with time, it is important to measure the potential and verify that each individual RE potential is stable in order to compare results. Such characterizations can be performed by measuring the open circuit potential (OCP) of all REs against a master RE or by any other suitable published method. REs can also be calibrated with a Pt WE to the reversible hydrogen electrode (RHE). Note that the potential for the RHE versus NHE is dependent on the pH (or more strictly, the hydrogen activity) of the electrolyte. To perform this calibration, slow current–potential scans (e.g., 1 mV/s) around the region expected for hydrogen evolution/oxidation in a H_2 saturated electrolyte are used and the point of intersection at the potential axis represents 0 V versus RHE. To limit potential drift between experiments, REs should always be stored in the solution recommended (or provided) by the supplier. Note that the filling solution as well as its concentration should be mentioned when reporting a potential measured against REs. Finally, although PEC electrode characterizations are performed at room temperature, temperature drift (e.g., heating by the lamp) may occur which can shift the potential of the reference electrode.

3.2.4 Choosing the Electrolyte

3.2.4.1 Fundamental Considerations

The electrolyte type (acidic, neutral, or basic) should be selected, so that the semiconductor of interest does not corrode when immersed in the solution. Some general guidance in electrolyte selection can be obtained from Pourbaix diagrams. As an example, the Pourbaix diagram of WO_3 is presented in Fig. 3.8.

WO_3 is stable below pH $= 2$ and at anodic potentials, and thus acidic electrolytes are often used. However, defining general electrolyte selection rules is a difficult task, since physical and chemical properties of semiconductor materials may vary depending on different deposition techniques. Some examples of

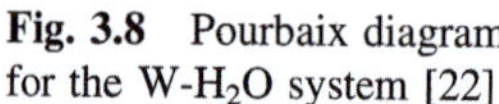

Fig. 3.8 Pourbaix diagram for the W-H$_2$O system [22]

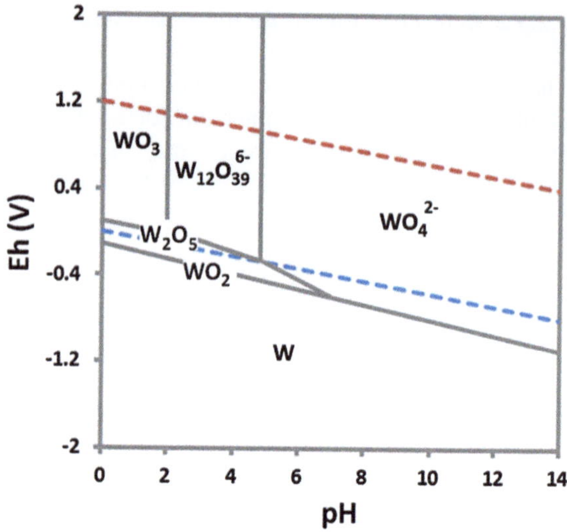

electrolyte choices for characterizing different semiconductors are presented in Table 3.2. If material stability is unknown, neutral electrolytes are a good starting point, although the kinetics for driving HER or OER is lower at neutral pH than in respective acidic or alkaline media. The electrolyte should not have absorption bands in the spectral window of the semiconductor's absorption, and known chemical interactions should be avoided (for example, carbonate buffers may chemisorb to some metal oxide surfaces). Finally, the electrolyte concentration should be sufficiently ionically conductive ($\sim$0.1 M or above) to ensure minimum solution resistance.

Table 3.2 A sample of electrolytes commonly employed in PEC electrode tests

Material	Conductivity type	Solution	Electrolyte concentration (M)	pH
GaInP$_2$	p-type	Sulfuric acid	3	−0.5
WO$_3$	n-type	Phosphoric acid	0.33	1.28
	n-type	Sulfuric acid	0.5	0.4
CuGaSe$_2$	p-type	Sulfuric acid	0.5	0.4
InGaN	n-type	Carbonate buffer		10
a-SiN	n-type	Carbonate buffer		10
Fe$_2$O$_3$	n-type	Sodium hydroxide	1	13.6
	n-type	Potassium sulfate	0.5	2
Cu$_2$O	p-type	Sodium hydroxide	0.1	13
	p-type	Sodium acetate	0.1	7.4
TiO$_2$	n-type	Any	0.1	All

3.2.4.2 Preparing the Electrolyte for PEC Electrode Testing

Electrolytes with specific concentrations can be prepared by diluting a concentrated solution with high purity deionized water (resistivity greater than 17–18 MΩ.cm to minimize contaminants). In a two-electrode measurement performed in a single compartment cell, the electrolyte should be sparged with an inert gas (such as argon or nitrogen) for 10 min prior to the experiment. This will remove oxygen dissolved in the solution that can be easily reduced and compete with the HER. During a three-electrode measurement, the solution should be sparged with H_2 gas for p-type electrodes and oxygen gas for n-type electrodes to establish a defined redox potential. In a double compartment cell, each compartment should be sparged with the respective gas that is being produced. If a large amount of hydrogen or oxygen is produced during the experiment (i.e., $j_{ph} > 1$ mA/cm^2), a small concentration of surfactant (e.g., Triton-X) [23] can be added to the electrolyte to ensure that the evolved gas bubbles remain small and do not stick to the electrode surface. Otherwise, these bubbles would reduce the electrochemically active surface area. Note that these organic surfactants may be participating in the reaction, for instance, by being selectively oxidized or reduced in place of O_2 or H_2 evolution. Stirring (e.g., using a magnetic stir bar) or other agitation can also be used to remove bubbles.

3.2.5 Connecting the Electrodes to the Potentiostat

In a three-electrode configuration, the potentiostat attains a specific potential difference between the WE and RE by varying the current between the WE and CE. When a two-electrode configuration is chosen, the potential difference between the WE and CE is maintained by varying the current flow between them. Source meters and potentiostats can come with two, three, four, or five electrode connections. Researchers should consult the manual for these devices to determine the proper terminal connections.

3.2.6 Device Testing Approaches

Several approaches can be used to perform two- or three-electrode measurements of PEC devices in configurations compatible with installation in reactor designs. In the open cell configuration (Fig. 3.9a), the WE (semiconductor of interest) with the active film facing up is placed along the inside edge of the container, close to the center of the edge. A CE is placed at one side of the WE. When a three-electrode configuration is required, the RE is placed as close as possible to the WE (a Luggin capillary tube was used in this case) to limit the effect of electrolyte resistance. This configuration could also be used vertically in an open beaker

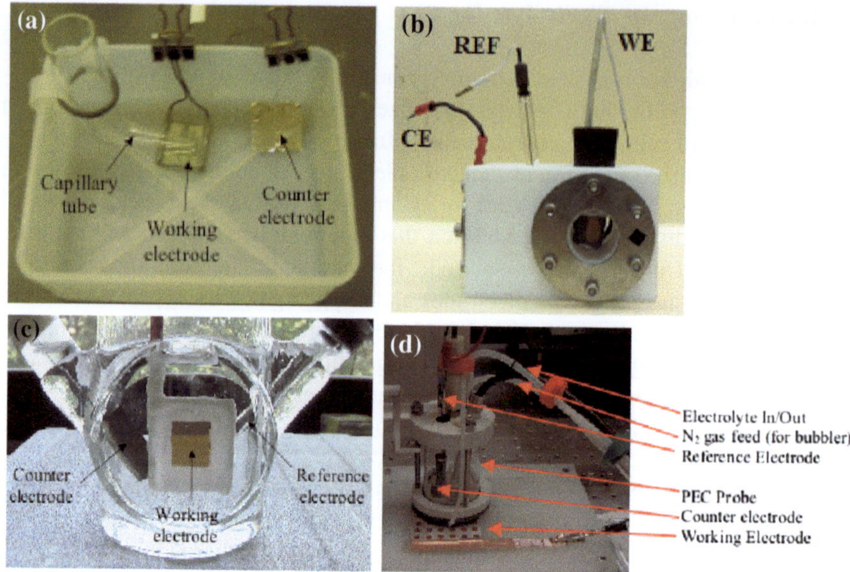

Fig. 3.9 Examples of electrochemical setups used for PEC device testing, including (**a**) open beaker with a Luggin capillary tube for the RE, (**b**) three-port Teflon cell, (**c**) three-port glass cell, and (**d**) robotic probe for combinatorial analysis

where light enters from the side of the container rather than from above. In the three-port glass cell configuration (Fig. 3.9b, c), each electrode is placed in a dedicated port. The position of the RE is tuned to minimize the distance from the WE. More advanced setups can be used to perform combinatorial analyses such as the arrangement presented in Fig. 3.9d. This system consists of a robotic probe head containing the CE, the RE, a gas diffuser, and an optical fiber bundle which is isolated from the electrolyte via a flat wall borosilicate tube. The probe is contacted to the sample and sealed using an O-ring, exposing the electrolyte and the illumination source to a constant geometric area.

3.3 Catalyst Surface Treatments

For efficient PEC solar conversion, good light absorption and carrier transport in the semiconductor bulk is essential. The next critical factor is the efficient extraction of carriers at the interfaces of both the hydrogen and oxygen evolving electrode surfaces. Some commonly employed surface modification techniques to catalytically enhance these surfaces are discussed here.

3.3.1 Principle of Surface Catalysis

Kinetic or activation overpotential is defined as the potential in excess of the thermodynamic equilibrium (E^0) potential of the half-reaction that is necessary to drive the reaction, and often plays an important role in determining overall reaction rates. A means of increasing the reaction rate is to specifically improve the kinetics of the surface reactions by the addition of catalyst materials which can reduce these overpotentials. In metal electrodes, the modification of the surface with a catalyst increases the current at a given applied potential. Similarly in photoelectrodes, addition of a catalyst can ensure efficient catalytic turnover of photogenerated minority charge carriers at the semiconductor–electrolyte interface.

Employing a three-electrode j–V measurement (see Section "Three-Electrode j–V and Photocurrent Onset") allows the catalysis at the working electrode (WE) to be distinguished independently of the reaction at the counter electrode (CE) and thus this measurement is extremely useful for catalyst development. This is because only the potential of the WE versus the reference electrode (RE) is of concern in a three-electrode configuration, and the overpotential required at the CE is of little consequence. However, two-electrode devices require consideration of the full cell voltage, and thus reducing the activation overpotential at both the WE and the CE by using efficient catalysts on both electrodes is necessary to achieve the highest possible short-circuit photocurrent in a PEC device. Additionally, a unique advantage of a single electrode absorber device is that its CE need not be illuminated, and is therefore not constrained to the same surface area as the WE. Thus, the CE can be oversized relative to the WE to reduce its current density (thereby lowering the overpotential), enabling the use of less efficient and lower cost catalysts.

3.3.2 Selecting the Catalyst

In general, the catalyst should first be selected according to the nature of the reaction at the surface of the PEC electrode. For the hydrogen evolution reaction (HER), metals such as Pt and Pd, as well as Ni and Co (in alkaline solutions) are very efficient. However, Pt and Pd are usually poor catalysts for the oxygen evolution reaction (OER). Oxides such as RuO_2 (or hydroxides: $RuO_2 \bullet xH_2O$), IrO_2, as well as Ni and Co mixed-metal oxides are better candidates for the OER [24–26]. The HER is the pertinent reaction on (1) the surface of a p-type photoelectrode and on (2) the surface of the CE which complements a n-type photoelectrode. Analogously, the OER is the pertinent reaction on (1) the surface of a n-type photoelectrode and on (2) the surface of the CE which complements a p-type electrode.

3.3.3 Morphological Considerations

The choice of metal or metal oxide is crucial for catalytic activity, but the manner in which it is deposited is also important. A conformal coating of catalyst material is often desirable as it may engender protecting effects [27] for the underlying electrode. However, catalyst films beyond a few monolayers may be undesirable for several reasons. First, many electrocatalysts may not be adequately transparent and can attenuate light reaching the semiconductor (shadowing), thereby reducing photocurrent. In addition, since most metals collect both charge carriers (electrons and holes) efficiently, the metal areas could act as recombination centers and significantly reduce the performance of the PEC device. Therefore, a discontinuous coverage of catalyst particles such that the interparticulate distance is less than the minority carrier diffusion length may help alleviate these limitations. This can be achieved by using synthetic approaches that disperse small catalyst islands over the surface. Ideally, the island size should be on the order of a few nanometers to provide sufficient active areas for the gas evolution to occur yet limit shadowing of the semiconductor surface. Several deposition methods are discussed in the following section.

3.3.4 Deposition Methods

The dimension and synthetic method of particulate catalysts on semiconductors have an impact on their performance. In this section, we will discuss several techniques used to fabricate particles and films of Pt and RuO_2, two materials widely used for PEC electrode catalysis.

3.3.4.1 Electrodeposition

Electrodeposition processes are simple, low-cost techniques that are broadly used in both bulk material deposition and surface treatment. In the case of catalyst deposition, the dimension of the islands may be controlled by adjusting process parameters such as applied potential, current density, or concentration of precursors. Pulsed periodic deposition profiles can also yield different morphologies. Scanning electron microscopy images of electrodeposited Pt and IrO_2 on n-Ta_3N_5 electrodes are shown in Fig. 3.10b, c, respectively. If the electrodeposition is performed under illumination (a.k.a. photodeposition), the catalyst will be preferentially deposited on active sites where photogenerated charge carriers are injected into the electrolyte, optimizing the usage of catalytic material.

Several experiments have demonstrated platinum catalyst particle deposition over p-type materials using cathodic processes [30–32]. A typical deposition can be performed using an electrolyte made of 0.01 M chloroplatinic acid (H_2PtCl_6) in

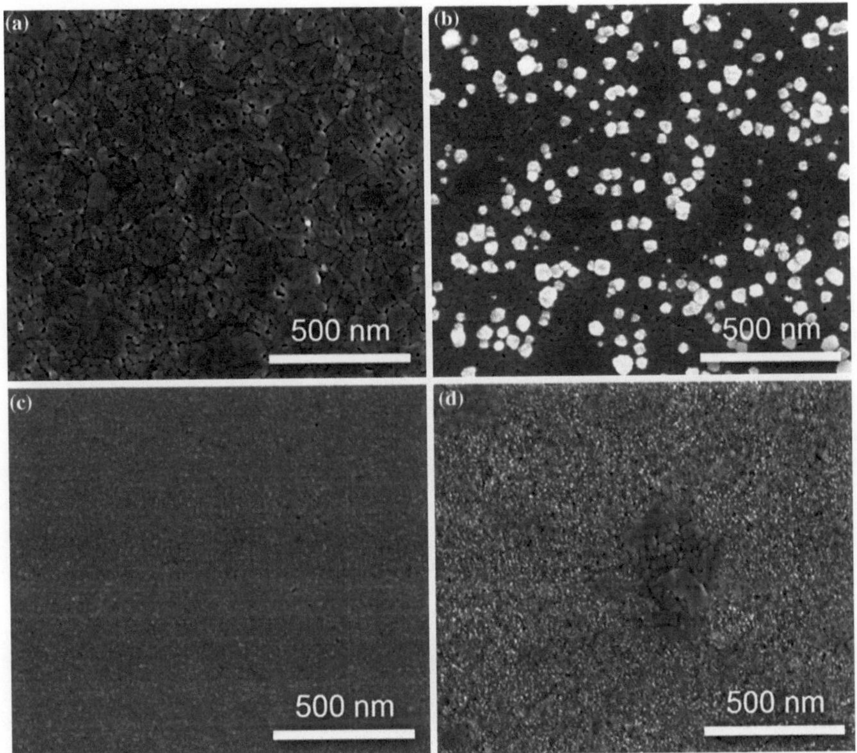

Fig. 3.10 Scanning electron microscopy images of (**a**) bare n-Ta$_3$N$_5$ electrode surface, (**b**) n-Ta$_3$N$_5$ electrode with electrodeposited Pt particles, (**c**) n-Ta$_3$N$_5$ electrode with IrO$_2$ electrodeposited from a colloidal IrO$_2$ solution prepared from an acidic condensation method [28], and (**d**) n-Ta$_3$N$_5$ electrode with vapor phase deposited RuO$_2$ from decomposition of RuO$_4$ [29]

2 M hydrochloric acid (HCl) under galvanic square wave conditions with a duty cycle of 0.25 s at -1 mA/cm^2 followed by 2 s at open circuit, this cycle is repeated until a total charge of 10 mC/cm^2 is reached. The addition of illumination may be necessary for p-type photocathodes in order to provide sufficient minority carrier electrons to drive cathodic electrodeposition. When a n-type semiconductor is used as a PEC electrode, typical cathodic electrodeposition (3–9 mA/cm^2) of RuO$_2$•xH$_2$O can be obtained from a 0.04 M aqueous acidic ruthenium chloride (RuCl$_3$•xH$_2$O) solution [33]. Anodic deposition of this material is also feasible by adding 0.1 M NaAcO [34].

3.3.4.2 Sputtering

PVD processes are usually employed to deposit continuous films with thicknesses ranging from 100 nm to a few microns. However, it is possible to form islands of material by interrupting the deposition process in the early stages of growth, before the initial clusters coalesce into a continuous film. Such process control can be achieved using low deposition rates. Both the substrate and the catalyst of interest may influence the size, the surface density, and even the electronic properties of the catalyst since different growth mechanisms can occur. These growth mechanisms may be two dimensional (Frank-van der Merwe growth mode), three dimensional (Volmer-Weber growth mode), or both (Stranski–Krastanov growth mode).

Platinum particles in the 3–4 nm range have been successfully deposited using a RF magnetron (100 W) sputtering process performed at room temperature [35]. Islands of RuO_2 have been observed in the initial stages of RuO_2 growth when using a reactive sputtering process (65 W RF power) at 200 °C [36].

3.3.4.3 Spray Deposition

The spray deposition method is typically best suited for depositing metal oxides such as RuO_2. The use of this method is limited to semiconductors that are stable upon heating (often in an oxygen-containing atmosphere) such as TiO_2 and Fe_2O_3. The method consists of spraying a dilute solution of the metal oxide precursor, for example $RuCl_3$ (1–10 mM in H_2O depending on desired thickness), onto a photoelectrode that is being heated on a hot plate (573 K) [37]. The deposition can be carried out with a spray nozzle, plastic atomizer, or custom-made nozzle. The deposition rate should not decrease the sample temperature substantially, and very short deposition times are required for an electrocatalyst since only a small amount of catalyst is desired. The deposition should be performed in a fume hood since the decomposition vapors of the metal oxide precursor are potentially hazardous.

3.3.4.4 Cryogenic Decomposition of Ruthenium Tetroxide

This method consists of the thermal decomposition of RuO_4 to RuO_2 on the surface of the photoanode. Calcination of the substrate at 300 °C for 30 min is suggested to obtain better contact with the RuO_2 nanoparticles, but it is not essential. This method is time consuming and requires a fume hood as well as acetone (or isopropanol) and dry ice to decrease the temperature of the RuO_4 mixed in pentane [38].

3.3.4.5 Vapor Deposition of Hydrous RuO_2 Thin Films

This process consists of the autocatalytic decomposition of RuO_4 to RuO_2 and is shown in Fig. 3.10d. RuO_4 can be used directly [39] if it can be formed in situ by mixtures [29, 40, 41] of an oxidizing agent and either hydrous ruthenium oxide ($RuO_2 \bullet xH_2O$) or hydrous ruthenium chloride ($RuCl_3 \bullet xH_2O$). The RuO_4 vapor is reduced to RuO_2 by autocatalytic decomposition on the substrate.

3.3.5 Electrical Characterization

Enhancing the catalysis at the surface of PEC electrodes results in a lower kinetic overpotential and an increase in photocurrent. The effectiveness of the catalysts after surface treatment can be determined by utilizing three-electrode $j–V$ measurements (see Section "Three-Electrode $j–V$ and Photocurrent Onset") as well as IPCE measurements (see Chapter "Incident Photon-to-Current Efficiency and Photocurrent Spectroscopy"). It may also useful to perform Mott-Schottky (see Section "Mott–Schottky") to determine any impacts these catalysts may have on the band structure (e.g., due to Fermi level pinning).

3.4 Spectral Standards

Since the spectral distribution and intensity of sunlight are so regionally and temporally variable, a standard reference spectrum has been established, so that results obtained from different researchers can be compared on equal footing. It is important to understand the spectral distribution as well as the intensity of terrestrial solar radiation when selecting which equipment combinations are most appropriate for simulating sunlight in a laboratory setting. It is critical for researchers to adhere to spectral standards guidelines in the measurement and reporting of all experimental results.

3.4.1 The AM 1.5 G Reference Spectrum

The Sun is modeled approximately as a 5900 K blackbody [42] radiator with a total irradiance of 1366.1 W/m^2 at the periphery of Earth's atmosphere [43]. Radiation passing through the atmosphere is attenuated by molecules (O_3, H_2O, CO_2) and suspended particulates that can scatter or absorb portions of the spectrum. The amount of attenuation depends on atmospheric parameters such as aerosol loading, humidity, and pressure. The resulting clear-sky spectrum reaching the surface is a function of many atmospheric parameters and ground reflections as well as the path length through the atmosphere. Because the path length is the easiest parameter to define, the air mass (AM) notation is commonly used to define spectra. The AM number represents the amount of atmosphere traversed. AM 0 is defined as the spectrum outside the Earth's atmosphere at 1 astronomical unit from the Sun, and AM 1 represents sunlight reaching the Earth's surface (at sea level) when the Sun is directly overhead. The AM number can vary between 1 and greater than 5 depending on the time of day, time of year, and latitude [44]. The AM number can be determined from the zenith angle of the Sun by the following expression:

$$AM\# = 1/\cos\theta_{zenith}$$

where θ_{zenith} is the angle between the vector toward the Sun and the local vertical vector normal to the surface [45]. The standard spectrum is chosen to be AM 1.5, because it represents approximately the average annual AM value available at locations within the continental US [44].

For AM 1.5, the standard spectrum is defined as the irradiance arriving on a flat plate tilted at 37° (which represents the average latitude for the contiguous 48 states) from horizontal toward the Sun. Because $\theta_{zenith} = 48.2°$ for AM 1.5 and the tilt angle is 37°, the angle of incidence is 11.2°. An intensity is specified, as well as the spectral distribution of illumination, with the accepted total irradiance (intensity) of AM 1.5 being 1000 W/m^2 (100 mW/cm^2). Under actual AM 1.5 conditions, the measured intensity would have a slightly lower magnitude, but the US Photovoltaics program scaled up the magnitude to achieve a round number and established the first standard AM 1.5 spectrum and intensity in 1977 [46].

In addition to the spectral distribution, the type of illumination is also relevant. Sunlight arriving on the surface of the Earth is a combination of direct and diffuse radiation. Diffuse radiation is the portion of illumination that has undergone forward scattering by aerosols and reflected light off the ground and cannot be focused or concentrated. Direct radiation is light collected as if looking through a tube toward the Sun and can be concentrated. Direct radiation also includes a component of diffuse radiation (specifically referred to as the circumsolar radiation). The current reference spectra assume the circumsolar component is 5.8° wide centered around the Sun [44]. The sum of direct and diffuse radiation is known as global and includes all light collected from a 180° field of view. For the AM 1.5 G standard spectrum (where G denotes global), 90 % of the intensity is due to direct radiation and 10 % is from diffuse radiation [47]. In actual measurements of a bright Sun on a cloudless but hazy day, the diffuse component of solar radiation can be responsible for up to 29 % of the total radiation [48]. Under total cloud cover, diffuse radiation accounts for 100 % of the intensity [49].

There are currently two tabulated spectral standards accepted by the photovoltaic (PV) community. One standard, ASTM G-173-03, maintained by the American Society of Testing and Materials, was adopted in 2008 and replaced the old standard ASTM G-159-98 or equivalently ASTM E-892. The reference spectrum used by the terrestrial community from 1985 to 2008 is ASTM G-159. This standard was generated with a computer model that is not available and cannot be recreated. The new reference spectrum was generated with a publicly available and well-documented computer model, and can be regenerated at using the same or different input parameters (http://www.nrel.gov/rredc/smarts) [43]. This new standard includes both global and direct spectra and can be found online at   http://www.astm.org/Standards/G173.htm   and   http://rredc.nrel.gov/solar/spectra/am1.5/. The ASTM G-173-03 standard is plotted in Fig. 3.11. The relevant international standard from the International Electrotechnical Commission is IEC 60904-3 edition 2 (http://www.iec.ch/). The IEC version renormalized the ASTM version, changed the number of digits, and added a small term from

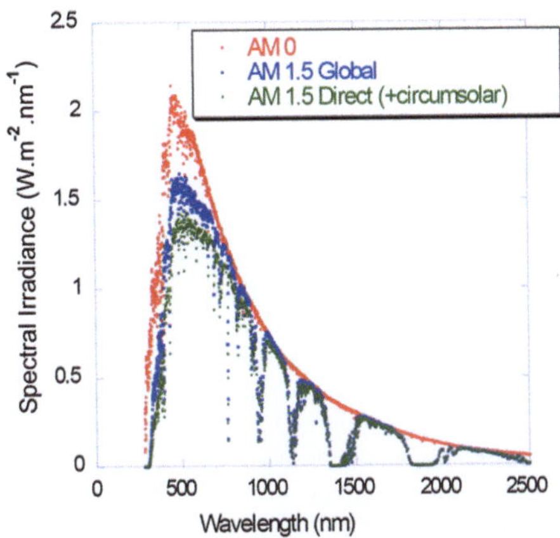

Fig. 3.11 AM 0, AM 1.5 global, and AM 1.5 direct spectra plotted using ASTM data

4,000 nm to infinity. These changes have a negligible impact on any results. Either standard is acceptable for use to combine with measured IPCE values to estimate efficiency under AM 1.5 G.

Integration of the AM 1.5 G spectral standard (for an idealized device where IPCE is 100 % for photon energies above the band gap) is a useful tool to determine the maximum short-circuit current density possible for any material. This is known as the optical limit, where every photon with energy above band gap is converted to an electron in the circuit. The maximum j_{sc} under AM 1.5 G can easily be converted to a maximum (upper-bound) solar-to-hydrogen (STH) conversion efficiency by multiplying by the free energy of the water splitting reaction at 25 °C (1.23 V) and dividing by the *entire* integrated AM 1.5 G spectrum (100 mW/cm^2), assuming 100 % Faradaic efficiency. The optical limit plot (Fig. 1.2 in Chapter "Introduction") can be used to verify plausibility of a measured efficiency. No correctly measured efficiency can exceed the optical limit.

Any laboratory photoconversion measurement that involves broadband illumination (white light) must take into account the light source. Arc sources put out considerably more intensity in the UV portion of the spectrum than filament sources. The advantages of filament lamps are lower cost and more stable output intensity. However, they also output a great deal of IR radiation, and thus passing the light through an actively cooled volume of water (commercial jacketed water filters that integrate with optical components are readily available) can prevent excessive heating of downstream components such as the electrolyte.

Commercial solar simulators are available that typically use arc sources and filters to mimic the AM 1.5 G spectrum, although usually they do not replicate it exactly. Consequently, some portions of the spectral output (especially particular emission lines from a particular source) have greater intensity, and some regions have lower intensity, as shown in Fig. 3.12.

Different classes of solar simulators are available; the classification is dependent on the spectral mismatch of the true AM 1.5 G spectrum. These devices are

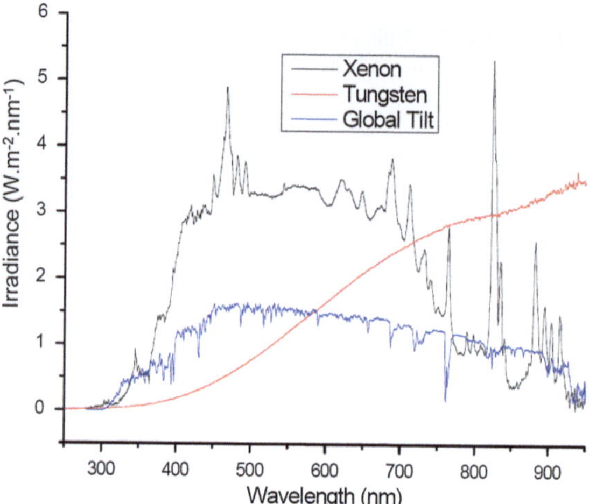

Fig. 3.12 Spectral output of a 1 kW Xe-arc lamp and a 250 W tungsten-halogen lamp as measured using a spectroradiometer consisting of a Si photodiode array calibrated against the NIST F-420 standard, compared to the AM 1.5 G standard (also known as global tilt). The output of the xenon lamp was passed through an IR absorbing water filter and a commercially available AM 1.5 filter which further attenuates UV and IR radiation

usually calibrated by measuring the short-circuit current density (j_{sc}) of a silicon (or other appropriate semiconductor) photodiode, which averages intensity differences because j_{sc} essentially integrates all photon flux (number of photons per second) above the band gap. Consequently, the calibration is generally a valid assumption for materials with a band gap near that of the detector (1.1 eV for silicon). However, large errors are introduced when wider band gap materials (such as those required for PEC water splitting) are tested on simulators calibrated with smaller band gap materials such as silicon (care should be taken and other standards should be considered). Although the integrated flux above 1.1 eV for a solar simulator may be equivalent to the reference spectrum, the two values diverge for different values of the band gap. This usually results in erroneously high efficiency values (sometimes greater than the optical limit) [47] if the flux of photons above the band gap for the solar simulator is greater than the reference AM 1.5 G spectrum. More detailed discussions on suitable light sources for emulating terrestrial solar radiation have been reported in the literature [44, 47].

3.4.2 Reference Cells

It is a nontrivial task to exactly replicate the "standard" solar spectrum in a laboratory setting. Fortunately, this onerous task is unnecessary, as there are methods for approximating real-world efficiencies from measurements taken using laboratory illumination sources that differ from the reference spectrum.

The use of a reference cell is one simple method that gives reliable real-world performance estimates based on efficiencies measured in the laboratory where illumination conditions do not match the solar spectrum. This is the accepted standard for determining PV efficiencies in the US [44]. A reference cell is a solid-state PV device that has had its j_{sc} calibrated under the AM 1.5 G spectrum. Short-circuit current is the parameter that is most sensitive to the spectral distribution of the light source [44]. It is important that the band gap of the reference cell be as close as possible to the material being tested, because the calibration cell can only correct for spectral mismatch if it absorbs the same portion of the spectrum as the test electrode. Otherwise, a reference cell can be fitted with short-pass filters, to mimic wider band gap devices, and calibrated. Calibration of reference cells can be performed by any of several recognized institutions (NREL, Fraunhofer, AIST, and others). Integration of the spectral response over the AM 1.5 G spectrum is another method that can be used to obtain j_{sc} and yield a reference cell that does not require outside testing.

To use a reference cell, the intensity of the light source is adjusted until the measured j_{sc} matches the calibrated value. The assumption is that although the spectral distribution of the light source may not match that of the standard reference spectrum, the integrated irradiances (as determined by j_{sc}) are equivalent. Stated another way, this approach matches the total flux above the band gap, and not necessarily the shape of the irradiance. This can introduce some error, especially near the band edge. Inexpensive light sources can be used to mimic AM 1.5 G with reasonable accuracy. One study found that the error in j_{sc} between simulated sunlight (filtered xenon arc lamps and dichroic filtered tungsten halogen lamps) and terrestrial sunlight was about 2 % for several reference cell-test cell combinations [50]. Due to the wavelength dependent response of semiconductor materials, efficiency measurements are sensitive to spectral variation. Therefore, it is essential when reporting water splitting efficiencies that the lamp type and intensity are reported as well as the band gap of the calibration device (e.g., reference cell, photodiode array, and thermopile). Recently, several groups have reported artifacts at the several percent level in measuring the light level in solar simulators. This artifact is maximum when the distance between the simulator and reference cell is short and there is an air gap between the reference cell window and the cell [51].

References

1. H.L. Wang, T. Deutsch, J.A. Turner, Direct water splitting under visible light with nanostructured hematite and WO_3 photoanodes and a $GaInP_2$ photocathode. J. Electrochem. Soc. **155**, F91–F96 (2008)
2. I. Matulionis, F. Zhu, J. Hu, T. Deutsch, A. Kunrath, E. Miller, B. Marsen, A. Madan, Development of a corrosion-resistant amorphous silicon carbide photoelectrode for solar-to-hydrogen photovoltaic/photoelectrochemical devices. Paper presented at Conference on Solar Hydrogen and Nanotechnology III (2008)
3. B. Marsen, B. Cole, E.L. Miller, Influence of sputter oxygen partial pressure on photoelectrochemical performance of tungsten oxide films. Sol. Energy Mater. Sol. Cells **91**, 1954–1958 (2007)

4. Y.-S. Hu, A. Kleiman-Shwarsctein, A.J. Forman, D. Hazen, J.-N. Park, E.W. McFarland, Pt-doped α-Fe$_2$O$_3$ thin films active for photoelectrochemical water splitting. Chem. Mat. **20**, 3803–3805 (2008)

5. A. Kleiman-Shwarsctein, Y.-S. Hu, A.J. Forman, G.D. Stucky, E.W. McFarland, Electrodeposition of α-Fe$_2$O$_3$ doped with Mo or Cr as photoanodes for photocatalytic water splitting. J. Phys. Chem. C **112**, 15900–15907 (2008)

6. W.M. Sachtler, G.J.H. Dorgelo, A.A. Holscher, Work function of gold. Surf. Sci. **5**, 221 (1966)

7. J. Westlinder, G. Sjoblom, J. Olsson, Variable work function in MOS capacitors utilizing nitrogen-controlled TiNx gate electrodes. Microelectron. Eng. **75**, 389–396 (2004)

8. N. Gaillard, M. Gros-Jean, D. Mariolle, F. Bertin, A. Bsiesy, Method to assess the grain crystallographic orientation with a submicronic spatial resolution using Kelvin probe force microscope. Appl. Phys. Lett. **89**, 154101 (2006)

9. M.D. Deal, J.D. Plummer, P.B. Griffin, *Silicon VLSI Technology Fundamentals, Practice and Modeling* (Prentice Hall, Upper Saddle River, 2000), p. 817

10. B. Marsen, B. Cole, E.L. Miller, Photoelectrolysis of water using thin copper gallium diselenide electrodes. Sol. Energy Mater. Sol. Cells **92**, 1054–1058 (2008)

11. N.S. Gaikwad, G. Waldner, A. Bruger, A. Belaidi, S.M. Chaqour, M. Neumann-Spallart, Photoelectrochemical characterization of semitransparent WO$_3$ films. J. Electrochem. Soc. **152**, G411–G416 (2005)

12. O. Khaselev, J.A. Turner, A monolithic photovoltaic–photoelectrochemical device for hydrogen production via water splitting. Science **280**, 425–427 (1998)

13. T.G. Deutsch, C.A. Koval, J.A. Turner, III–V nitride epilayers for photoelectrochemical water splitting: GaPN and GaAsPN. J. Phys. Chem. B **110**, 25297–25307 (2006)

14. A. Kay, I. Cesar, M. Gratzel, New benchmark for water photooxidation by nanostructured α-Fe$_2$O$_3$ films. J. Am. Chem. Soc. **128**, 15714–15721 (2006)

15. W. Siripala, A. Ivanovskaya, T.F. Jaramillo, S.H. Baeck, E.W. McFarland, A Cu$_2$O/TiO$_2$ heterojunction thin film cathode for photoelectrocatalysis. Sol. Energy Mater. Sol. Cells **77**, 229–237 (2003)

16. J.-N. Nian, C.-C. Hu, H. Teng, Electrodeposited p-type Cu$_2$O for H$_2$ evolution from photoelectrolysis of water under visible light illumination. Int. J. Hydrog. Energy **33**, 2897–2903 (2008)

17. T.F. Jaramillo, S.H. Baeck, A. Kleiman-Shwarsctein, K.S. Choi, G.D. Stucky, E.W. McFarland, Automated electrochemical synthesis and photoelectrochemical characterization of Zn$_{1-x}$Co$_x$O thin films for solar hydrogen production. J. Comb. Chem. **7**, 264–271 (2005)

18. L.J. Minggu, W.R.W. Daud, M.B. Kassim, An overview of photocells and photoreactors for photoelectrochemical water splitting. Int. J. Hydrog. Energy **35**, 5233–5244 (2010)

19. S. Haussener, C. Xiang, J.M. Spurgeon, S. Ardo, N.S. Lewis, A.Z. Weber, Modeling, simulation, and design criteria for photoelectrochemical water splitting systems. Energy Environ. Sci. **5**, 9922–9935 (2012)

20. Z. Chen, T.F. Jaramillo, T.G. Deutsch, A. Kleiman-Shwarsctein, A.J. Forman, N. Gaillard, R. Garland, K. Takanabe, C. Heske, M. Sunkara, E.W. McFarland, K. Domen, E.L. Miller, J.A. Turner, H.N. Dinh, Accelerating materials development for photoelectrochemical hydrogen production: Standards for methods, definitions, and reporting protocols. J. Mater. Res. **25**, 3–16 (2010)

21. P. Vanysek, *CRC Handbook of Chemistry and Physics. Electrochemical Series*, vol 78, (CRC Press, Boca Raton, 1997), pp. 8-20–8-33

22. M. Pourbaix, *Atlas of Electrochemical Equilibria in Aqueous Solutions* (NACE, Houston, 1974)

23. O. Khaselev, J.A. Turner, A monolithic photovoltaic–photoelectrochemical device for hydrogen production via water splitting. Science **280**, 425–427 (1998)

24. A.K.M.F. Kibria, S.A. Tarafdar, Electrochemical studies of a nickel–copper electrode for the oxygen evolution reaction (OER). Int. J. Hydrog. Energy **27**, 879–884 (2002)

25. M.F. Kibria, M.S. Mridha, Electrochemical studies of the nickel electrode for the oxygen evolution reaction. Int. J. Hydrog. Energy **21**, 179–182 (1996)
26. E.L. Miller, R.E. Rocheleau, Electrochemical behavior of reactively sputtered iron-doped nickel oxide. J. Electrochem. Soc. **144**, 3072–3077 (1997)
27. Z. Chen, D. Cummins, B.N. Reinecke, E. Clark, M.K. Sunkara, T.F. Jaramillo, Core–shell MoO_3–MoS_2 nanowires for hydrogen evolution: a functional design for electrocatalytic materials. Nano Lett. **11**, 4168–4175 (2011)
28. Y. Zhao, E.A. Hernandez-Pagan, N.M. Vargas-Barbosa, J.L. Dysart, T.E. Mallouk, A High Yield Synthesis of Ligand-Free Iridium Oxide Nanoparticles with High Electrocatalytic Activity. The Journal of Physical Chemistry Letters **2**, 402–406 (2011)
29. A. Kleiman-Schwarsctein, A.B. Laursen, F. Cavalca, W. Tang, S. Dahl, I. Chorkendorff, A general route for RuO_2 deposition on metal oxides from RuO_4. Chem. Commun. **48**, 967–969 (2011)
30. M. Szklarczyk, J.O.M. Bockris, Photoelectrochemical evolution of hydrogen on p-indium phosphide. J. Phys. Chem. **88**, 5241–5242 (1984)
31. R.N. Dominey, N.S. Lewis, J.A. Bruce, D.C. Bookbinder, M.S. Wrighton, Improvement of photoelectrochemical hydrogen generation by surface modification of p-type silicon semiconductor photocathodes. J. Am. Chem. Soc. **104**, 467–482 (1982)
32. R.C. Kainthla, B. Zelenay, J.O.M. Bockris, Significant efficiency increase in self-driven photoelectrochemical cell for water photoelectrolysis. J. Electrochem. Soc. **134**, 841–845 (1987)
33. B.-O. Park, C.D. Lokhande, H.-S. Park, K.-D. Jung, O.-S. Joo, Cathodic electrodeposition of RuO_2 thin films from Ru(III)Cl_3 solution. Mater. Chem. Phys. **87**, 59–66 (2004)
34. C.-C. Hu, M.-J. Liu, K.-H. Chang, Anodic deposition of hydrous ruthenium oxide for supercapacitors. J. Power Sources **163**, 1126–1131 (2007)
35. M. Alvisi, G. Galtieri, L. Giorgi, R. Giorgi, E. Serra, M.A. Signore, Sputter deposition of Pt nanoclusters and thin films on PEM fuel cell electrodes. Surf. Coat. Technol. **200**, 1325–1329 (2005)
36. W.-T. Lee, D.-S. Tsai, Y.-M. Chen, Y.-S. Huang, W.-H. Chung, Area-selectively sputtering the RuO_2 nanorods array. Appl. Surf. Sci. **254**, 6915–6921 (2008)
37. T.P. Gujar, V.R. Shinde, C.D. Lokhande, W.-Y. Kim, K.-D. Jung, O.-S. Joo, Spray deposited amorphous RuO_2 for an effective use in electrochemical supercapacitor. Electrochem. Commun. **9**, 504–510 (2007)
38. J.V. Ryan, A.D. Berry, M.L. Anderson, J.W. Long, R.M. Stroud, V.M. Cepak, V.M. Browning, D.R. Rolison, C.I. Merzbacher, Electronic connection to the interior of a mesoporous insulator with nanowires of crystalline RuO_2. Nature Mater. **406**, 169–172 (2000)
39. K.S. Lyons, D.R. Rolison, Selective deposition of hydrous ruthenium oxide thin films (2003)
40. K.E. Swider-Lyons, C.T. Love, D.R. Rolison, Selective vapor deposition of hydrous RuO_2 thin films. J. Electrochem. Soc. **152**, C158–C162 (2005)
41. Z. Yuan, R.J. Puddephatt, M. Slayer, Low-temperature chemical vapor deposition of ruthenium dioxide from ruthenium tetroxide: A simple approach to high-purity RuO_2 films Chem. Mat. **5**, 908–910 (1993)
42. D.R. Myers, K. Emery, C. Gueymard, Revising and validating spectral irradiance reference standards for photovoltaic performance evaluation. J. Sol. Energy Eng. **126**, 567–574 (2004)
43. American Society for Testing Materials, Standard for Solar Constant and Air Mass Zero Solar Spectral Irradiance Tables, Standard ASTM E490-00a, West Conshocken, PA (2006)
44. R.J. Matson, K.A. Emery, R.E. Bird, Terrestrial solar spectra, solar simulation and solar cell short-circuit current calibration: A review. Solar Cells **11**, 105–145 (1984)
45. M.A. Green, *Solar Cells: Operating Principles, Technology and System Applications* (Prentice Hall, Englewood Cliffs, 1998)
46. Terrestrial Photovoltaic Measurement Procedures, National Aeronautics and Space Administration, Technical Report TM 73702 (1977)

47. A.B. Murphy, P.R.F. Barnes, L.K. Randeniya, I.C. Plumb, I.E. Grey, M.D. Horne, J.A. Glasscock, Efficiency of solar water splitting using semiconductor electrodes. Int. J. Hydrog. Energy **31**, 1999–2017 (2006)
48. K.W. Boer, The solar spectrum at typical clear weather days. Sol. Energy **19**, 525–538 (1977)
49. J.R. Bolton, D.O. Hall, Photochemical conversion and storage of solar energy. Ann. Rev. Energy **4**, 353–401 (1979)
50. K.A. Emery, Solar simulators and I-V measurement methods. Solar Cells **18**, 251–260 (1986)
51. D. Romang, J. Meier, R. Adelhelm, U. Kroll, Reference solar cell reflections in solar simulators, in *Proceedings of 26th European Photovoltaic Solar Energy Conference and Exhibition*, 3AV.1.64 (2011)

Chapter 4
PEC Characterization Flowchart

The goal of PEC materials development is to design a material system that has the potential to satisfy most, if not all, of the requirements for cost-effective PEC hydrogen production. Figure 4.1 presents a recommended flowchart for the characterization of candidate PEC materials. The key knowledge gained as well as limitations of the different characterization techniques, highlighted in Table 4.1, are described in detail in following sections of this document. A material that can survive the rigorous testing set forth in this flowchart will be a particularly promising candidate for incorporation into an industrially deployable device for PEC hydrogen production. This organized approach to PEC characterization is intended to streamline the process for material screening so that discovery of promising candidates occurs at a faster and more orderly pace.

Reference

1. Z. Chen, T.F. Jaramillo, T.G. Deutsch, A. Kleiman-Shwarsctein, A.J. Forman, N. Gaillard, R. Garland, K. Takanabe, C. Heske, M. Sunkara, E.W. McFarland, K. Domen, E.L. Miller, J.A. Turner, H.N. Dinh, Accelerating materials development for photoelectrochemical hydrogen production: standards for methods, definitions, and reporting protocols. J. Mater. Res. **25**, 3–16 (2010)

Z. Chen et al., *Photoelectrochemical Water Splitting,*
SpringerBriefs in Energy, DOI: 10.1007/978-1-4614-8298-7_4,
© The Author(s) 2013

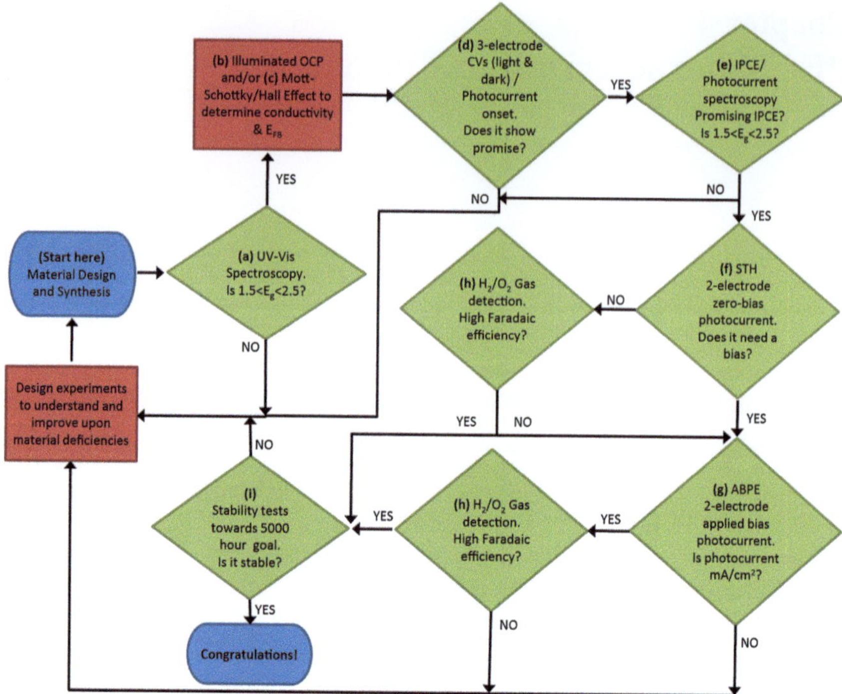

Fig. 4.1 Recommended PEC characterization flow chart for a single absorber material [1]

Table 4.1 List of key PEC characterization techniques, the knowledge gained from them, and their limitations

Technique	Knowledge gained	Limitations
(a) UV–Vis Spectroscopy	• Optical bulk band gap • Direct or indirect, allowed or forbidden transition	• Interpretation of data prone to subjective error and is not necessarily conclusive • Great absorbers do not necessarily make high efficiency PEC cells • Derived band gap for the bulk is not necessarily valid for the surface
(b) (Illuminated Open Circuit Potential)	• Flat-band potential (E_{fb}) • Conductivity type	• Possible photocorrosion at OCP • Samples with high recombination rate can prevent band flattening
(c) Mott-Schottky/Hall Effect	• Flat-band potential (E_{fb}) in Mott-Schottky • Conductivity type • Doping density • Density of charge carriers	• Complicated interpretation of data from non-ideal samples in Mott-Schottky • Samples should be amenable to deposition on non-conductive substrates for Hall effect measurements
(d) Three-Electrode Cyclic Voltammograms (Light and Dark)/Photocurrent Onset)	• Saturated photocurrent density • Voltage range of photocurrent generation • Conductivity type • Flat band potential	• Catalysts may be needed to minimize overpotential errors in flat band potential measurement • Possible sample corrosion
(e) IPCE/Photocurrent Spectroscopy	• Wavelength-dependent incident photon-to-current efficiency • Band gap • Absorbed photon-to-current efficiency	• High variability in data if the lamp source is not accurately measured • Possible photocorrosion or side reactions, which will lead to false (higher) IPCE values
(f) 2-electrode Zero-Bias Photocurrent	• Solar-to-hydrogen (STH) efficiency	• Possible photocorrosion or side reactions, which will lead to a false (higher) STH efficiency value if η_F is not accurate
(g) 2-electrode Applied Bias Photocurrent)	• Applied bias photon-to-current efficiency (ABPE)	• Possible photocorrosion or side reactions, which will lead to a false (higher) ABPE value
(h) H_2/O_2 Gas Detection)	• Faradaic efficiency for water splitting	• Quantitative detection requires perfect sealing of reaction vessel
(i) Stability Tests	• Lifetime of material	• Instability of illumination source over time (if not continuously measured)

Chapter 5
UV-Vis Spectroscopy

5.1 Knowledge Gained from UV-Vis Spectroscopy

In a UV-Vis (ultraviolet-visible light) spectroscopic measurement, light absorption as a function of wavelength provides information about electronic transitions occurring in the material. The fraction of light transmitted is described by the Beer-Lambert law, which states that the fraction of the light measured after interaction with the sample (I, usually measured as transmittance or reflectance) versus the incident intensity (I_0) is dependent on the path length of light through the sample (l), the absorption cross section (σ) of the transition, and the difference in the population of the initial state (N_1) and final state (N_2) of the initial (E_1) and final (E_2) electronic energy levels.

$$\frac{I}{I_0} = e^{-\sigma(N_1 - N_2)l} \tag{5.1}$$

This is often written in a form referred to simply as Beer's Law:

$$A = \varepsilon c l = -\log_{10}\left(\frac{I}{I_0}\right) \tag{5.2}$$

where A is the absorbance, ε is the molar absorptivity coefficient of the material, c is the concentration of the absorbing species, and l is the path length of light through the sample. Note that the measured intensity of a real system can be convoluted by transitions that occur across multiple initial and final states.

The absorbance A can be normalized to the path length l of the light through the material (e.g., the thickness of a film), producing the absorption coefficient α [1]:

$$\alpha\left(\mathrm{cm}^{-1}\right) = \frac{\ln(10) \times A}{l(\mathrm{cm})} \tag{5.3}$$

For semiconductors, UV-Vis spectroscopy offers a convenient method of estimating the optical band gap, since it probes electronic transitions between the

Z. Chen et al., *Photoelectrochemical Water Splitting*,
SpringerBriefs in Energy, DOI: 10.1007/978-1-4614-8298-7_5,
© The Author(s) 2013

valence band and the conduction band. The optical band gap is not necessarily equal to the electronic band gap, which is defined as the energy difference between the valence band maximum (VBM) and the conduction band minimum (CBM), though it is often approximated as such because there are few, if any, convenient methods for measuring the electronic band gap. Exciton binding energies, d–d transitions, phonon absorptions and emissions, and excitations to or from defect bands and color centers can complicate interpretation of UV-Vis spectra; nevertheless, an estimation of the optical band gap is obtainable. Furthermore, UV-Vis theoretically allows for the characterization of this electronic transition as either a direct or indirect transition and also whether it is allowed or forbidden, although actual determination is not necessarily so straightforward, as will be illustrated below. A direct transition is described as a two-particle interaction between an electron and a photon, whereas an indirect transition is described as a three-particle interaction (photon, electron, and phonon) to ensure momentum conservation. A transition is allowed or forbidden depending on the dipole selection rules associated with the system. The shape of the UV-Vis absorption spectrum can, in principle, distinguish between these transitions by analysis of Tauc plots. The "efficiency" of the photon absorption process occurring within a sample, $A\%$ (related to $\eta_{e-/h+}$ described in Chapter "Introduction"), formally known as absorptance (i.e., different from absorbance), is defined as the fraction of photons absorbed per photons impinging on the sample:

$$A_\% = \eta_{e-/h+} = 1 - \frac{I}{I_0} = 1 - 10^{-A}. \tag{5.4}$$

5.2 Limitations of UV-Vis Spectroscopy

The UV-Vis measurement is relatively straightforward, and the data obtained is generally reproducible from lab to lab despite differences in lamp sources, spectrometers, experimental configuration, etc. However, there are some experimental pitfalls that need to be avoided. Additionally, deriving a band gap value from a UV-Vis measurement can be prone to error. Interpretation is often limited by the shape of the absorption spectrum and the ability of the user to estimate the line tangent to the slope of the absorption data to assess its onset. This procedure is highly subjective and can result in significant error if not interpreted correctly.

A primary source of error in a UV-Vis measurement often arises from reflection or scattering that may occur at the surface and interfaces of the sample. The following equations represent the various components that incident light can split into after interaction with the sample.

$$I_0 = A_\% + T + R_S + R_d + S \tag{5.5}$$

where $A_\%$ is absorptance, T is transmitted light, R_s is specularly reflected light, R_d is diffusely reflected (back-scattered) light, and S accounts for other forms of refracted light that are redirected off-axis (forward-scattered). Figures 5.1 and 5.3 depict these various components in a transmission and diffuse reflectance measurement, respectively. It is important to keep in mind that the parameter being measured (typically either T or R_d) is not solely affected by $A_\%$.

In a transmission experiment, the measured I is represented by the following equation:

$$I = T = I_0 - (A_\% + R_S + R_d + S) \tag{5.6}$$

Whereas in a reflectance measurement (which employs an integrating sphere), the measured I is represented by R_d, as follows:

$$I = R_d = I_0 - (A_\% + T + R_S + S) \tag{5.7}$$

These other effects (R_s, S, and T or R_d depending on configuration) can decrease the amount of light that reaches the detector (I) and produce seemingly higher absorption values. This can result in nonzero baselines or sloped baselines that need to be taken into account when analyzing spectra. To minimize reflection or refraction during a transmission experiment, the user should ensure that the sample is not tilted but sits normal to the path of incident light. Scattering effects can be minimized by placing the sample as close as possible to the detector in a transmission experiment. In a highly scattering sample, reflectance or absorption approaches that utilize an integrating sphere may provide better signal.

In the case of very thin samples in transmission measurements, Fabry–Perot interference fringes can arise as a result of sample-support interactions which can manifest as an oscillatory signal that convolutes the data and further complicates the data interpretation process. Though proper modeling of these features can produce the true absorbance while also revealing additional information about the sample, a workaround is to utilize a diffuse reflectance or absorption configuration as detailed below.

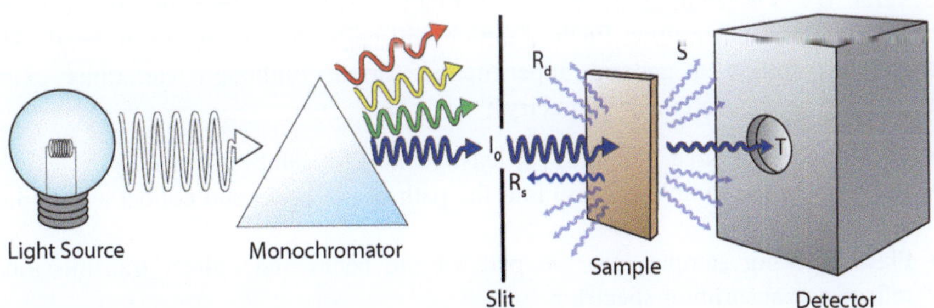

Fig. 5.1 Single beam UV-Vis transmission configuration with the monochromator placed before the sample

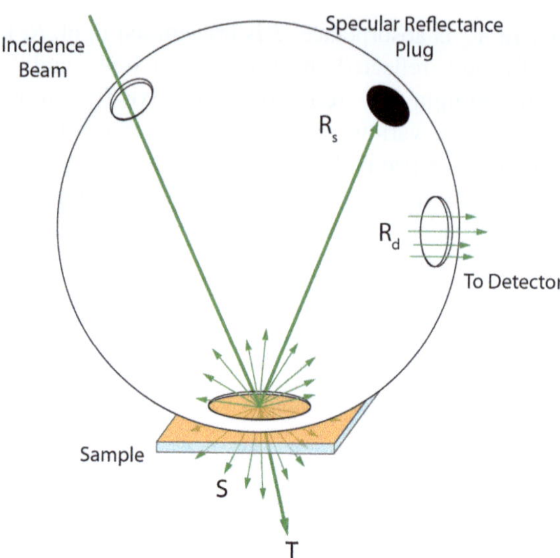

Fig. 5.2 Diffuse reflectance configuration using an integrating sphere with a specular reflectance plug

Pitfalls of the experimental procedure often come from an improperly positioned sample that can lead to stray travel of light, as mentioned above, or from measurements performed before the lamp has had proper time to warm up, resulting in drift of the light source. Improper shielding of the sampling chamber from ambient lighting can also contribute to the background signal and decrease the signal-to-noise ratio. Lastly, high-order diffraction peaks that arise from using a monochromator can lead to inaccurate measurements if they are not removed using long pass filters.

5.3 Method for Performing UV-Vis Spectroscopic Measurements

Several UV-Vis configurations are available, including transmission, diffuse reflectance, and absorption. Each of which will be further discussed in detail. All techniques follow a general experimental format (although variations exist depending on measurement geometry):

- Turn on lamp source and allow at least 15 min for lamp to warm up
- Place reference sample (if any) into the path of the beam and collect a baseline scan
- Place working sample into the path of the beam and collect transmission/reflectance/absorption spectrum
- Calculate the absorption coefficient (if thickness is known) and create Tauc plots
- Fit the spectrum to determine size and type of band gap.

There is little to no preparation time involved in a UV-Vis spectroscopic measurement. The amount of time required to conduct the experiment is minimal and mostly determined by the time required for the lamp to warm up. Sampling time itself will vary depending on the speed and range of the scan but generally takes no more than a few minutes. Analysis time is similarly expedient, requiring no more than a few minutes to plot the data in a spreadsheet program.

5.3.1 Experimental Parameters

The experimentalist must first decide between using a transmission, diffuse reflectance, or absorption configuration. These modes will be discussed in detail below. In general, transmission mode is used for samples that have some degree of transparency. Often these materials are thin films supported on substrates transparent to visible light. Opaque samples, such as materials supported on metallic substrates, cannot be used in transmission mode since the spectrometer would receive no signal. As such, opaque samples must utilize a diffuse reflectance or absorption configuration. In a diffuse reflectance configuration, use of a light absorbing plug (see below) can minimize contributions of specular reflectance to signal at the detector. Diffuse reflectance is also the configuration of choice for samples in particulate form, such as powders or grains, since the larger acceptance angle of the integrating sphere can potentially provide greater signal compared to a measurement made in transmission mode where much of the signal will be scattered off-axis.

5.3.2 Transmission UV-Vis

In transmission configuration, the user places the sample of interest, hereby referred to as the working sample, in the path of a collimated beam of light. Samples must have some degree of transparency. The light impinges upon the working sample and is partially absorbed at characteristic wavelengths corresponding to electronic transitions in the sample. A spectrometer collects the transmitted light and compares the output against a baseline measurement that is referenced as 100 % transmission (I_0). The reference measurement must take into account the absorbance by any material support, such as a cuvette holder or a glass slide. Transmission reference measurements can be accomplished using either a one-beam or two-beam setup.

In a one-beam transmission configuration, as shown in Fig. 5.1, the user first places the reference sample (e.g., an empty cuvette or a clean support free of the absorber material of interest) in the path of the beam and performs a baseline scan. Afterward, the user replaces the reference sample with the working sample for measurement. The drawback to this method is the potential for drift and other fluctuations in the beam to occur over time, especially during warm up. Therefore, it is best to perform a reference scan immediately prior to performing the scan on the working sample to minimize time for any fluctuations that may occur. Two-beam instrument configurations (not shown here) are also available which can maintain a dynamic baseline to further minimize fluctuations.

5.3.3 Diffuse Reflectance UV-Vis

In a diffuse reflectance configuration, the spectrometer measures the diffusely reflected light, rather than the transmitted light, from a sample. Samples best suited to this configuration include powders or films. A typical configuration for a diffuse reflectance measurement involves the use of an integrating sphere to capture all photons that are reflected (in all directions) from the sample, as shown in Fig. 5.2. A typical integrating sphere has an input port connected to the light source, an output port connected to a signal meter that collects the diffusely reflected light, and an aperture against which the working or reference samples can be placed for measurement. The following discussion pertains to setups utilizing this type of integrating sphere. The inside of an integrating sphere is covered with a highly reflective material such as polytetrafluoroethylene (PTFE) or Ba_2SO_4, which are effective over a large wavelength region of interest. This material also serves as a nearly ideal Lambertian scatterer by distributing the light uniformly throughout the entire surface of the integrating sphere.

Two types of reflection can occur: specular and diffuse. Specular reflection occurs when the incident beam of radiation strikes a sample and reflects at an angle that is equal to the angle of incidence. Specularly reflected light has not undergone an absorption process, and thus contains little to no information regarding electronic states within the material. In diffuse reflectance, the incident beam penetrates the sample surface, is partially absorbed, and a fraction of its photons is reemitted (reflected) at various nonincident angles. During diffuse reflectance measurements, specular reflectance will increase noise, decrease the accuracy of the measurement, and can contribute to spurious peaks in the data. An integrating sphere can contain a specular reflectance sink (or "plug") that minimizes this contribution. For powders, dilution in a nonabsorbing matrix can further increase diffuse reflectance while minimizing specular reflectance. Typical nonabsorbing matrix materials include KBr, KCl, and Ba_2SO_4.

A diffuse reflectance measurement begins with the collection of a reference scan. For samples deposited on reflective substrates, such as metallic molybdenum, the bare metal serves as a reference to account for any absorption in the metal itself. Samples deposited onto transparent substrates, such as a transparent conducting oxide deposited on glass (e.g., indium-tin oxide, or ITO), require the use of a white diffuse reflectance standard. This standard is often made from Ba_2SO_4 or PTFE based material similar to, if not the same as, that used to coat the interior surface of the integrating sphere. The user places the standard against the open aperture, or sampling port, of the integrating sphere and the spectrometer collects a baseline which is then used as the reference "spectrum" for 100 % reflection. Some instruments also support the ability to collect a dark scan to minimize contributions from stray illumination. For samples supported on a transmissive support, it is advisable to fit a bare support free of the absorbing material of interest between the aperture and the reflectance standard to also account for any absorbance or scattering due to the support itself. However, this does not guarantee accurate accounting of all scattered light, especially in the case of highly scattering samples supported on thick substrates that can very effectively channel light along the plane of the sample.

Following the reference and dark scans, the working sample is placed against the aperture of the integrating sphere. The spectrometer then determines the amount of light reflected by the sample by comparison against the reference standard.

5.3.4 Absorption UV-Vis

A sample can also be placed directly within an integrating sphere as shown in Fig. 5.3. Ideally, this allows for collection of transmitted, reflected, and scattered light, leaving the measured difference relative to the reference attributable to absorptance alone, i.e.,

$$I = (I_0 - A_\%) = T + R_d + R_S + S \qquad (5.8)$$

This setup can minimize errors due to losses from scattering and reflectance. However, since the light will experience multiple passes through the sample due to

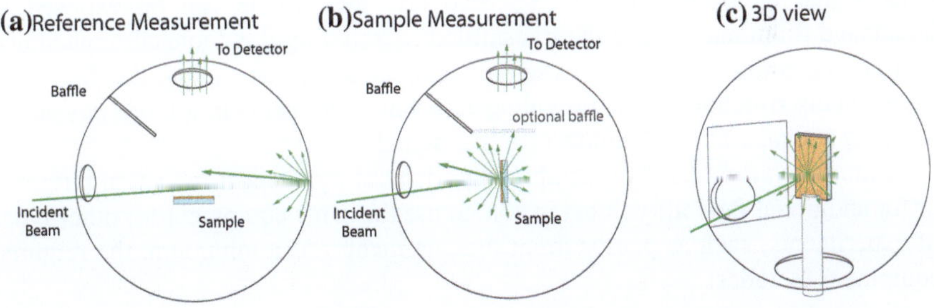

Fig. 5.3 Top-down view of an integrating sphere which fully encases the sample, which is placed inside using a holder that extends from the top or bottom through a third port. (**a**) Reference measurement in which the sample is placed out of the path of the primary beam. (**b**) Measurement of the sample itself in which the beam strikes the sample. (**c**) 3D view of experimental setup showing the sample placed in a vertical holder. The incidence beam enters at a slight angle to ensure that specularly reflected light is not lost by exiting back through the entrance port. A baffle prevents incident light from directly traveling to the detector

reflection from the walls of the integrating sphere, obtaining an absorption coefficient may be inaccurate unless steps are taken to account for the additional interactions of light. To account for additional absorptance from 2nd, 3rd, 4th... nth passes of light through the sample, one can take a reference measurement in which the sample is placed within the integrating sphere, but tilted out of the way of the primary beam path as shown in Fig. 5.3b. In this configuration, the primary beam first diffuses throughout the sphere before interacting with the sample, and this baseline signal can be approximated as due to 2nd, 3rd, 4th... nth passes. The remaining signal measured with the sample in the path of the primary incident beam can then be approximated as due to the 1st pass. The addition of baffles can minimize errors due to scattered, reflected, or transmitted light from the sample that would otherwise reach the detector without interacting with the sphere surface. The actual measured absorptance can be prone to inaccuracy, however, since the sample geometry is necessarily different relative to the reference measurement. Scattering or reflectance in preferential orientations may result in lost signal (e.g., through open ports of the sphere) or may simply produce a different path of beam travel between the two configurations that introduces error into the comparison.

5.3.5 *Required Equipment for UV-Vis Measurements*

Commercial UV-Vis spectrometers are widely available in both transmission and diffuse reflectance configurations. Depending on the configuration, these machines can offer an extremely wide wavelength window over which measurements can be performed, extending from UV to IR radiation. Generally, a monochromator is positioned before or after the sample. If the source light is monochromated, the sample is illuminated with a single wavelength at any given moment, and the transmitted/reflected light is detected with the aid of a photomultiplier tube or other signal enhancing device. Alternatively, the sample can be exposed to broadband illumination, and the transmitted/reflected light is monochromated and sent into detector. In the latter case, a photodiode array can capture the data at all wavelengths simultaneously, providing faster measurements in a more physically compact package at the expense of lower signal.

A modular setup can be assembled in a laboratory for cost reduction purposes or for additional flexibility in being able to use the same equipment for other types of experiments, such as photocurrent measurements. At a minimum, the required equipment includes:

- Light source
- Monochromator with long pass filters
- Integrating sphere (for reflectance and absorption measurements)
- Suitable detector such as a power meter
- Various focusing lenses and optical fibers.

The light source can come in the form of arc lamps (mercury, tungsten-halogen, xenon) or tunable dye lasers. Care must be taken to maintain sample integrity during broad spectral range illumination. For example, xenon lamps produce large intensity in the infrared spectral region, which can locally heat and possibly damage a sample if an infrared absorbing filter (such as a water column) is not used. High-energy UV photons may also damage certain samples.

The choice of lamp is often dictated by the wavelength range desired for a given experiment. For work that primarily requires a large amount of UV radiation in the range of 160–400 nm, deuterium lamps are preferable. For larger ranges of 200–2500 nm, xenon and mercury lamps are suitable. Many off-the-shelf instruments contain multiple lamps to adequately cover the full spectrum of interest.

Long pass filters eliminate spurious bands that arise from harmonics in the separation of light in the monochromator. For single-grating monochromators, band pass filters eliminate broadband stray light that increases rapidly for wavelengths shorter than 400 nm. Additionally, if the monochromator is set to output light with a wavelength of 700 nm, light at wavelengths of 350–175 nm will also be emitted. Therefore, a long pass filter that eliminates the lower wavelength peaks would be required to properly control the output of the monochromator.

Integrating spheres are commercially available from multiple vendors. For a given light input, a smaller sphere will be brighter than a larger sphere since the internal surface area is smaller. However, light throughput can be negatively affected if a sphere is too small, since the presence of input, output, and sampling ports distort the ideal spherical geometry and decrease the hemispherical reflectance. Many integrating spheres have an optimum diameter of 2–4 inches with total area of the ports making up approximately 5 % of the total internal surface area [2]. However, depending on sample size, larger spheres may be desirable to minimize errors from scattering or from having too many passes of the light through the sample, particularly if the sample is placed within the integrating sphere itself.

5.4 Analysis of Band Gap Energies from UV-Vis Spectra

An ideal UV-Vis spectrum collected for a defect-free direct band gap semiconductor exhibits almost no absorption for photons with energies below the band gap and a sharp increase in absorption for photons with energies above the band gap. Since spectra are typically reported in units corresponding to the wavelength of light rather than its energy, the conversion between conventional wavelength (nm) and band gap energy (eV) units is achieved by:

$$h\nu\,(\text{eV}) = \frac{1239.8\ (\text{eV} \times \text{nm})}{\lambda\,(\text{nm})} \tag{5.9}$$

The band gap in the absorption spectrum corresponds to the point at which absorption begins to increase from the baseline, since this indicates the minimum amount of energy required for a photon to excite an electron across the band gap and thus be absorbed in the semiconductor material. Real spectra exhibit a non-linear increase in absorption that partly reflects the local density of states at the conduction band minimum and valence band maximum, as well as other factors such as defect states [3].

In a transmission experiment, the instrument software may use Eq. (5.2) to directly calculate absorbance from the measured intensity. However, the measured intensity is affected not only by absorptance, but by reflectance and scattering as well (i.e., $A_\% \neq I_0\text{–}T$). These effects are often related to the morphology of each sample (e.g., a sample with a rough surface will introduce significant light scattering that decreases the amount of light reaching the detector and consequently increases the perceived absorbance). These effects often manifest in the form of a nonzero baseline. One way to correct these effects is to shift all the data so that the data point with the lowest absorbance value corresponds to zero absorbance. This method makes the assumption that any reflectance and scattering effects are wavelength independent. It is important to realize that this assumption is not always valid and can introduce error in the data analysis.

A detailed band gap analysis involves plotting and fitting the absorption data to the expected trendlines for direct and indirect band gap semiconductors. Ideally, the absorbance A is first normalized to the path length l of the light through the material to produce the absorption coefficient α as per Eq. (5.3). Values of $\alpha > 10^4$ cm^{-1} often obey the following relation presented by Tauc and supported by Davis and Mott [4, 5]:

$$\alpha h v \propto \left(h v - E_g \right)^{1/n} \tag{5.10}$$

where n can take on values of 3, 2, 3/2, or 1/2, corresponding to indirect (forbidden), indirect (allowed), direct (forbidden), and direct (allowed) transitions, respectively, [1], [6–8]. These so-called Tauc plots [9–11] of $(\alpha h v)^n$ versus $h v$ yield the value of the band gap when extrapolated to the baseline, as summarized in Table 5.1.

For values of $\alpha < 10^4$ cm^{-1}, an exponential tail exists for many materials that cannot be modeled by Eq. (5.10) [3, 5, 12–14]. As such, one should attempt to fit to data points at energies that are greater than the energy required to achieve $\alpha > 10^4$ cm^{-1}. Fitting a tangent to a point within this tail will underestimate the band gap of the material. However, the value of 10^4 cm^{-1} used to distinguish regions of the absorption coefficient is not a strict cutoff and will vary from one system to another. In many cases during materials development, the exact path length may be unknown. It is still possible to ascertain a band gap value without first normalizing A to α, but the researcher should be aware of potential error in underestimating the onset as just described. The presence of mid-band gap (e.g., defect or dopant) states will also absorb light of lower energies, which can complicate data interpretation.

In the case of a diffuse reflectance measurement, in which one measures $I = R_d$, the Kubelka–Munk radiative transfer model can be employed to extract α [2, 15, 16].

$$f(R) = \frac{(1-R)^2}{2R} = \frac{\alpha}{s} \tag{5.11}$$

where $f(R)$ is the Kubelka–Munk function and s is the scattering coefficient. If the scattering coefficient is assumed to be wavelength independent, then $f(R)$ is proportional to α and the Tauc plots can be made using $f(R)$ in place of α [17–20]. However, since it is not possible to accurately plot the value of α without knowing the scattering coefficient, care must be taken to extrapolate the band gap from higher values of $f(R)$. Otherwise, extrapolation from the region of the exponential tail can lead to underestimation, as noted above. It is important to note that the assumption of wavelength independency for s can lead to error due to effects such as Fabry–Perot interference as mentioned previously; a more rigorous analysis of obtaining the absorption coefficient from diffuse reflectance is described by Murphy and should be dutifully employed by researchers whenever possible [21, 22].

To estimate the nature and value of the band gap, the experimentally derived absorption curve can be plotted according to Table 5.1. As an example, the absorbance of an electrodeposited polycrystalline Cu_2O sample measured using a transmission configuration is shown in Fig. 5.4. The absorbance data was first

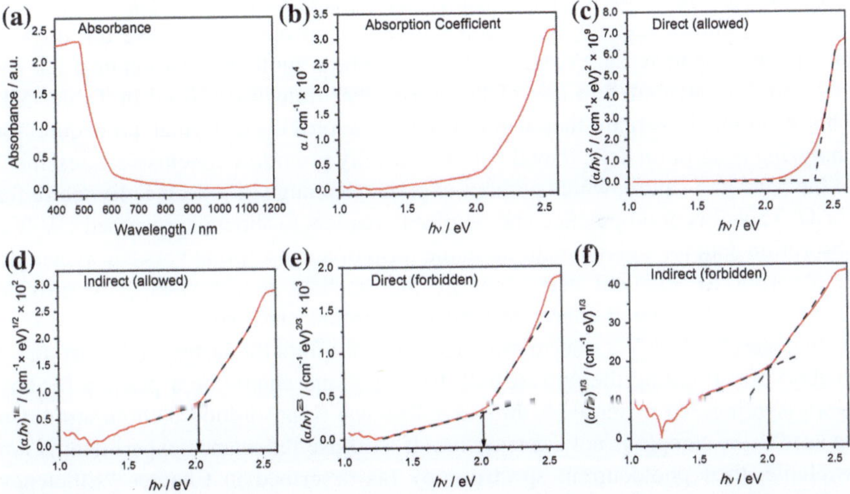

Fig. 5.4 (a) Absorbance data from a 1.7 μm film of electrodeposited polycrystalline Cu_2O plotted as (b) Normalized absorption coefficient vs. energy and in (c) Allowed direct (d) Allowed indirect (e) Forbidden direct, and (f) Forbidden indirect band gap Tauc plots [26]. Plot (c) Suggests an allowed direct transition with a band gap of approximately 2.4 eV, as reported by others [27–30]. Plots (d–f) suggest transitions near 2.0 eV. Cu_2O literature supports both allowed indirect [31, 32] and forbidden direct [33, 34] transitions. Allowed direct transitions near 2.0 eV have also been observed [35–37], and signal from an absorption tail below 2.0 eV has previously been attributed to copper ion vacancies and free carriers [23–25]

Table 5.1 Tauc plots and their respective transition types

Plot	Transition
$(\alpha h v)^2$ vs hv	Direct (allowed)
$(\alpha h v)^{2/3}$ vs hv	Direct (forbidden)
$(\alpha h v)^{1/2}$ vs hv	Indirect (allowed)
$(\alpha h v)^{1/3}$ vs hv	Indirect (forbidden)

shifted such that the minimum value was set to zero in order to account for any wavelength-independent reflectance and scattering. The absorbance was then analyzed using Eq. (5.3) for its absorption coefficient in plot (b) and in the form of Tauc plots in (c–f). For plots (d–f), a tangent is first drawn to the baseline at low energies (from 1.3–2.0 eV in this case). This step may not be necessary for plots that have little to no baseline, as is often the case with direct (allowed) band gap Tauc plots such as plot (c). Drawing a baseline in such a manner only accounts for reflection or scattering that exhibits a linear dependence with photon energies, and does not take into account nonlinear effects (e.g., plasmonic scattering). In this particular example, Cu_2O has contributions to its baseline from an absorption tail apparent in plots (c–e) that has previously been attributed to absorption from copper ion vacancies and free carriers [23–25]. Second, a line tangent to the slope in the linear region of the absorption onset is drawn. The intersection of the two lines corresponds to the best estimate for the energy of the band gap.

Plot (c) shows that this particular Cu_2O sample appears to have an allowed direct band gap near 2.4 eV, while plots (d–f) show another transition near 2.0 eV that could be attributed to any of the three other transitions listed in Table 5.1. This example illustrates that while Tauc plots provide a formal procedure for analyzing absorption data, they do not necessarily provide a conclusive assessment of the band gap nature, which perhaps explains the range of reports in literature for Cu_2O. Thus it is good practice and helpful to readers to show unprocessed UV-Vis absorption data for a given material in the form of an absorption versus wavelength or absorption coefficient versus energy plot such as in Fig. 5.4a, b to provide insight as to the photon energy at which absorption onset occurs.

In general, a UV-Vis transmission experiment offers the fastest and most direct method of estimating the optical bulk band gap and should be a priority for any newly synthesized material. A diffuse reflectance or absorption configuration can be used if the sample is not transmissive. If a diffuse reflectance experiment is not available, then photocurrent spectroscopy (as described in Chapter "Efficiency definitions in the field of PEC") with extremely facile redox couples can be performed, though errors in this method may arise from poor charge carrier mobilities or lifetimes and from slow kinetics at the sample-electrolyte interface.

References

1. S.M. Sze and K.K. Ng: Physics of semiconductor devices (Wiley, New York, 2006)
2. Integrating spheres–introduction and theory. *Comprehensive Catalog of FTIR Accessories and Supplies* (2005)
3. J.D. Dow, D. Redfield, Electroabsorption in semiconductors: The excitonic absorption edge. Phys. Rev. B **1**, 3358–3371 (1970)
4. E.A. Davis, N.F. Mott, Conduction in non-crystalline systems V. Conductivity, optical absorption and photoconductivity in amorphous semiconductors. Phil. Mag. **22**, 903–922 (1970)
5. D.L. Wood and J. Tauc, Weak absorption tails in amorphous semiconductors. *Phys. Rev. B, 5*, 3144–3151 (1972)
6. M. Anwar, C.A. Hogarth, Optical properties of amorphous thin films of MoO_3 deposited by vacuum evaporation. Phys. Status Solidi **109**, 469 (1988)
7. K. Santra, C.K. Sarkar, M.K. Mukherjee, B. Ghosh, Copper oxide thin films grown by plasma evaporation method. Thin Solid Films **213**, 226–229 (1992)
8. F.P. Koffyberg, K. Dwight, A. Wold, Interband transitions of semiconducting oxides determined from photoelectrolysis spectra. Solid State Commun. **30**, 433–437 (1979)
9. R.J. Elliott, Intensity of optical absorption by excitons. Phys. Rev. **108**, 1384–1389 (1957)
10. J. Tauc, Grigorov.R and A. Vancu, Optical properties and electronic structure of amorphous germanium. *J. Phys. Soc. Jpn. S 21*, 123 (1966)
11. J. Tauc, A. Menth and D.L. Wood, optical and magnetic investigations of localized states in semiconducting glasses. *Phys. Rev. Lett.* 25, 749–752 (1970)
12. F. Urbach, The long-wavelength edge of photographic sensitivity and of the electronic absorption of solids. Phys. Rev. **92**, 1324 (1953)
13. R.A. Street, N.F. Mott, States in the gap in glassy semiconductors. Phys. Rev. Lett. **35**, 1293–1296 (1975)
14. J. Tauc, A. Menth, D.L. Wood, Optical and magnetic investigations of the localized states in semiconducting glasses. Phys. Rev. Lett. **25**, 749–752 (1970)
15. Y.I. Kim, S.J. Atherton, E.S. Brigham, T.E. Mallouk, Sensitized layered metal oxide semiconductor particles for photochemical hydrogen evolution from nonsacrificial electron donors. J. Phys. Chem. **97**, 11802–11810 (1993)
16. P. Kubelka, New contributions to th eoptics of intensely light-scattering materials. Part I. J. Opt. Soc. Am. **38**, 449–457 (1948)
17. A.P. Finlayson, V.N. Tsaneva, L. Lyons, M. Clark, B.A. Glowacki, Evaluation of Bi-W-oxides for visible light photocatalysis. Phys. Status Solidi **203**, 327–335 (2006)
18. N. Kislov, S.S. Srinivasan, Y. Emirov, E.K. Stefanakos, Optical absorption red and blue shifts in ZnFe2O4 nanoparticles. Mater. Sci. Eng. B **153**, 70–77 (2008)
19. E.S. Brigham, C.S. Weisbecker, W.E. Rudzinski, T.E. Mallouk, Stabilization of intrazeolitic cadmium telluride nanoclusters by ion exchange. Chem. Mater. **8**, 2121–2127 (1996)
20. D.G. Barton, M. Shtein, R.D. Wilson, S.L. Soled, E. Iglesia, Structure and electronic properties of solid acids based on tungsten oxide nanostructures. J. Phys. Chem. B **103**, 630–640 (1999)
21. A.B. Murphy, Band-gap determination from diffuse reflectance measurements of semiconductor films, and application to photoelectrochemical water splitting. Sol. Energy Mater. Sol. Cells **91**, 1326–1337 (2007)
22. A.B. Murphy, Optical properties of an optically rough coating from inversion of diffuse reflectance measurements. Appl. Opt. **46**, 3133–3143 (2007)
23. V.F. Drobny, D.L. Pulfrey, Properties of reactively-sputtered copper-oxide thin-films. Thin.Solid.Films **61**, 89–98 (1979)
24. A.E. Rakhshani, J. Varghese, Optical-absorption coefficient and thickness measurement of electrodeposited films of Cu2O. Phys. Status Solidi A-Appl. Res. **101**, 479–486 (1987)

25. H. Wieder, A.W. Czanderna, Optical properties of copper oxide films. J. Appl. Phys. **37**, 184–187 (1966)
26. Z. Chen, T.F. Jaramillo, T.G. Deutsch, A. Kleiman-Shwarsctein, A.J. Forman, N. Gaillard, R. Garland, K. Takanabe, C. Heske, M. Sunkara, E.W. McFarland, K. Domen, E.L. Miller, J.A. Turner, H.N. Dinh, Accelerating materials development for photoelectrochemical hydrogen production: Standards for methods, definitions, and reporting protocols. J. Mater. Res. **25**, 3–16 (2010)
27. X. Mathew, N.R. Mathews, P.J. Sebastian, Temperature dependence of the optical transitions in electrodeposited Cu_2O thin films. Sol. Energy Mater. Sol. Cells **70**, 277–286 (2001)
28. B. Balamurugan, B.R. Mehta, D.K. Avasthi, F. Singh, A.K. Arora, M. Rajalakshmi, G. Raghavan, A.K. Tyagi, S.M. Shivaprasad, Modifying the nanocrystalline characteristics-structure, size, and surface states of copper oxide thin films by high-energy heavy-ion irradiation. J. Appl. Phys. **92**, 3304–3310 (2002)
29. J.F. Pierson, A. Thobor-Keck, A. Billard, Cuprite, paramelaconite and tenorite films deposited by reactive magnetron sputtering. Appl. Surf. Sci. **210**, 359–367 (2003)
30. N.A.M. Shanid, M.A. Khadar, Evolution of nanostructure, phase transition and band gap tailoring in oxidized Cu thin films. Thin Solid Films **516**, 6245–6252 (2008)
31. T. Kosugi, S. Kaneko, Novel spray-pyrolysis deposition of cuprous oxide thin films. J. Am. Ceram. Soc. **81**, 3117–3124 (1998)
32. A.E. Rakhshani, Preparation, characteristics and photovoltaic properties of cuprous-oxide–a review. Solid-State. Electron. **29**, 7–17 (1986)
33. K. Santra, C.K. Sarkar, M.K. Mukherjee, B. Ghosh, Copper-oxide thin-films grown by plasma evaporation method. Thin Solid Films **213**, 226–229 (1992)
34. P.E. de Jongh, D. Vanmaekelbergh, J.J. Kelly, Cu_2O: Electrodeposition and characterization. Chem. Mat. **11**, 3512–3517 (1999)
35. A.S. Reddy, G.V. Rao, S. Uthanna, P.S. Reddy, Structural and optical studies on do reactive magnetron sputtered Cu_2O films. Mater. Lett. **60**, 1617–1621 (2006)
36. T. Mahalingam, J.S.P. Chitra, J.P. Chu, H.Moon, H.J. Kwon and Y.D. Kim, Photoelectrochemical solar cell studies on electroplated cuprous oxide thin films. *J. Mater. Sci.-Mater. Electron.* **17**, 519–523 (2006)
37. W. Siripala, L. Perera, K.T.L. DeSilva, J. Jayanetti, I.M. Dharmadasa, Study of annealing effects of cuprous oxide grown by electrodeposition technique. Sol. Energy Mater. Sol. Cells **44**, 251–260 (1996)

Chapter 6
Flat-Band Potential Techniques

It is important to determine the conductivity and flat-band potential (E_{fb}) of a photoelectrode before carrying out any photoelectrochemical experiments. These properties help to elucidate the band structure of a semiconductor which ultimately determines its ability to drive efficient water splitting. Photoanodes (n-type conductivity) drive the oxygen evolution reaction (OER) at the electrode–electrolyte interface, while photocathodes (p-type conductivity) drive the hydrogen evolution reaction (HER). The conductivity type is determined from the direction of the shift in the open circuit potential upon illumination. Illuminating the electrode surface will shift the Fermi level of the bulk (measured potential) towards more anodic potentials for a p-type material and towards more cathodic potentials for a n-type material. The conductivity type is also used to determine the potential ranges for three-electrode j–V measurements (see section "Three-Electrode j–V and Photocurrent Onset") and type of suitable electrolyte solutions (see section "Cell Setup and Connections for Three- and Two-Electrode Configurations") used for the electrochemical analyses.

The three different techniques that can estimate the E_{fb} are: illuminated OCP (Section "Illuminated Open-Circuit Potential (OCP)"), Mott–Schottky (Section "Mott–Schottky") and photocurrent onset (Section "Three-Electrode j–V and Photocurrent Onset"). The E_{fb} should be independent of the technique used to determine it. Due to the inherent shortcomings of each technique, there is often a lack of agreement of the values determined by the various analyses. Researchers should be aware of these limitations in interpreting results.

6.1 Illuminated Open-Circuit Potential (OCP)

6.1.1 Knowledge Gained

The illuminated OCP can be used to estimate the E_{fb} if the above-band gap illumination ($hv \geq E_g$) is sufficiently intense to completely remove pre-existing band bending at the surface and the material does not have very fast carrier

Z. Chen et al., *Photoelectrochemical Water Splitting*,
SpringerBriefs in Energy, DOI: 10.1007/978-1-4614-8298-7_6,

recombination rates. The required intensity depends on the spectral characteristics of the light source used (see section "Spectral Standards"), as well as on the band gap and defect density of the material under study. To be sure, a plot of OCP versus light intensity provides the best method of estimating E_{fb}.

There are several ways to measure E_{fb}, ranging from simple to complex. In practice it is difficult to precisely quantify E_{fb}, as there are errors inherent to each type of measurement. Illuminated OCP is the easiest measurement for determining E_{fb}, and provides an accurate estimate in the case of a bright lamp and a high-quality (low recombination) sample. Above-gap illumination leads to photogenerated electron–hole pairs which are separated under the influence of the electric field within the space charge region. The minority charge carriers move towards the semiconductor-electrolyte interface while the remaining majority charge carriers generate a field opposing the electric field of the space charge layer. With sufficiently intense illumination, this will continue until the opposing fields equal in magnitude. Under these conditions, the measured voltage between the semiconductor and a reference electrode (RE) approximates the E_{fb} versus the reference electrode (RE) potential (the E_{fb} is thus defined to be the field-free potential), as shown in Fig. 6.1d. The magnitude of the difference in the OCP of the semiconductor in the dark (Fig. 6.1b) versus under illumination (Fig. 6.1c, d), as measured versus the RE, is the photovoltage, V_{ph}. Note, however, that the V_{ph} measured at open circuit may not exactly reflect the V_{ph} under operating conditions since the latter must also include the kinetic overpotential for the reaction of interest.

6.1.2 Limits of Experiments

On a short timescale (seconds to minutes), illuminated OCP analysis is a non-destructive technique for most materials. However, extended periods of illumination at open-circuit conditions may lead to corrosion of the photoelectrode surface [2]. Therefore, it is best to minimize the time of exposure to high-intensity illumination. Intense illumination can heat the solution (especially if the infrared radiation is not pre-filtered) at the electrode surface, which can cause a slow drift of the measured potential over the course of seconds or hours, depending on the rate of heating. In addition, drifts in the potential response may also be the result of corrosion processes or slow adsorption of cations or anions in solution to the semiconductor surface. A more rapid photovoltage response is often desirable and can be indicative of a better material.

The determination of E_{fb} values by OCP can also be complicated by materials that have a high density of defect sites which can serve as recombination centers. If the sample has a high carrier recombination rate (again due to material defects), then this effect prevents the creation of a compensating electric field, and hence higher light intensities are required to achieve flat-band conditions. However, if the lamp is not sufficiently intense, it may not be possible to completely flatten the bands with illumination. As a consequence of incomplete band flattening, the measured potential will underestimate the E_{fb}. Failure to fully flatten the bands at the semiconductor/electrolyte interface causes measured E_{fb} to be more cathodic than the true E_{fb} for p-type and more anodic than the true E_{fb} for n-type systems.

6.1.3 Method

A high impedance digital multimeter or a potentiostat can be used to determine the conductivity type of the material in a three-electrode electrochemical cell. If using a multimeter, connect the $+V$ (typically red) to the semiconductor and COM lead (typically black) to a reference electrode. If using a potentiostat, connect the WE lead to the semiconductor and the RE and CE leads to the appropriate electrodes (see section "Cell Setup and Connections for Three- and Two-Electrode Configurations"). Exclude all stray light from the setup by using any necessary covers and note the OCP with the light off. Illuminate the semiconductor with sufficiently intense illumination and note the OCP with the light on (Fig. 6.2).

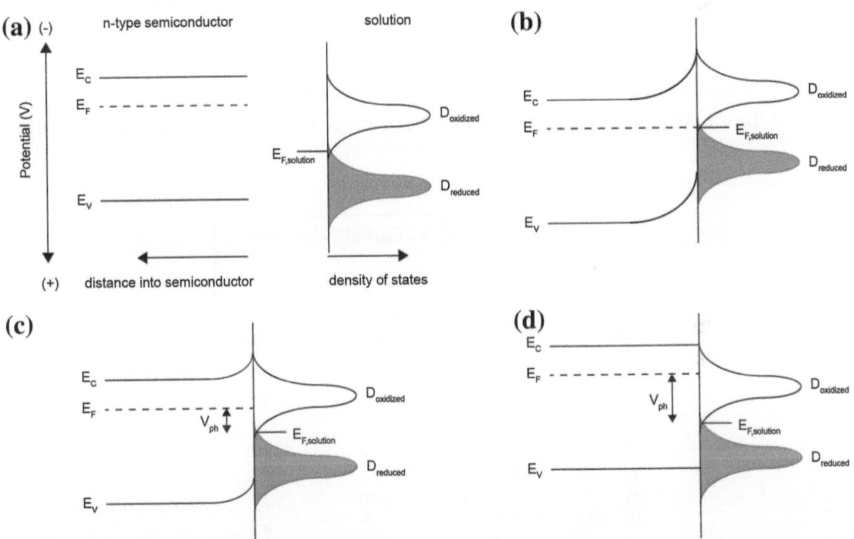

Fig. 6.1   Band diagrams of a n-type semiconductor (**a**) prior to contact with the electrolyte solution (assuming no defects or surface state charges), (**b**) in contact with the solution in absence of illumination, (**c**) in contact with the solution in the presence of moderate illumination, and (**d**) in contact with the solution in the presence of intense illumination and at the E_{fb}. Illustrated are the conduction band (E_C), Fermi level (E_F), and valence band (E_V) of the semiconductor. Also shown are the Gaussian distribution of the redox states in the solution, shown as the density of states of *oxidized* ($D_{oxidized}$) and *reduced* ($D_{reduced}$) species along with the corresponding Fermi level ($E_{F,solution}$), as described in more detail elsewhere [1]

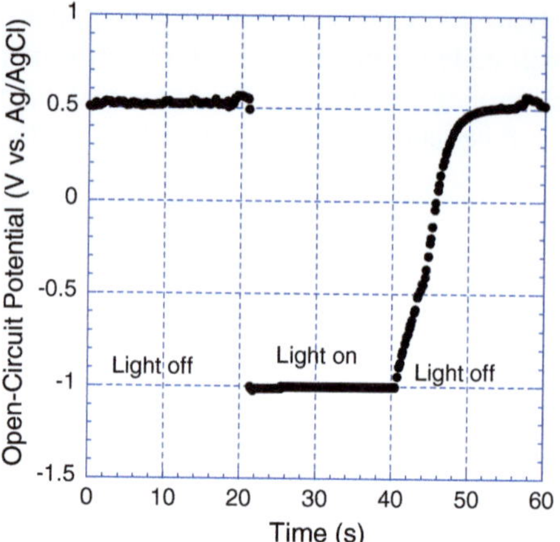

Fig. 6.2 Open-circuit potential of a GaInP$_2$ electrode in pH = 2 buffer with light off and light on. The negative shift in OCP upon illumination indicates the material possesses n-type conductivity, and the difference in OCP in the dark (+0.5 V vs. Ag/AgCl) and the OCP under illumination (−1 V vs. Ag/AgCl) indicates a V_{ph} of 1.5 V

One way to determine if the light intensity is sufficient to flatten the bands is to plot OCP versus illumination intensity (Fig. 6.3). For a material that does not experience significant Fermi level pinning [3, 4], a sufficiently intense illumination would saturate the OCP, where additional illumination has a negligible effect on OCP. It may take a few to tens of seconds or more (this is materials dependent) to

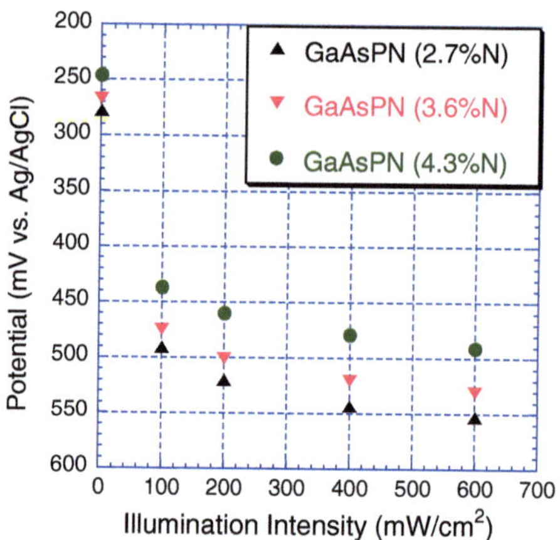

Fig. 6.3 Open-circuit potential (OCP) of three p-GaAsPN compositions [5, 6] ($E_g \sim 1.8$ eV) plotted versus illumination intensity [7]. The light source used for these measurements was a fiber optic illuminator equipped with a 150-Watt tungsten bulb, and the intensity was measured using an optical power meter with a thermopile detector. At 0.6 W/cm^2 (approximating six Suns of intensity), OCP is nearing saturation for this material/lamp combination

reach a steady state potential upon illumination. Unstable values for the OCP in the dark or under illumination could be due to electrolyte heating, adsorption of electrolyte species at the surface, and/or corrosion reactions, as described above.

6.1.4 Analysis

To determine the conductivity type, note the direction of potential shift with illumination. If OCP moves **P**ositive (towards more anodic potentials) with illumination, the material is p-type. If OCP moves **N**egative (towards more cathodic potentials), the material is n-type. If the potential did not change with illumination, there may be an issue with electrode fabrication/contacts, the material may be photo-inactive under these conditions, or the material may not be viable for PEC applications. If no response to illumination is observed, it is doubtful that the material, as mounted, will respond to any other photoelectrochemical characterization techniques. However, the researcher may still wish to perform CV scans as described in section "Three-Electrode j–V and Photocurrent Onset" to completely rule out photoactivity of the material.

6.1.5 Open-Circuit Potential and pH

It is common practice to measure E_{fb} across a range of pH values. The E_{fb} of many semiconductors follows a Nernstian relationship as a function of pH (Fig. 6.4). This is due to the dynamic equilibrium of the surface hydride/hydroxyl terminations that generally cause the band edges to move by -59 mV/pH [8]. A common way to analyze data is to plot the measured E_{fb} values along with the OER and HER potentials versus pH. By plotting the E_{fb} data along with the reaction potentials (which also exhibit Nernstian behavior), the ability to shift the band edges of the semiconductor with pH can be evaluated.

For a single photoelectrode material to drive spontaneous water splitting, a necessary (although by itself insufficient) requirement is that the E_{fb} of a n-type semiconductor must be negative of the hydrogen evolution potential (0 V vs. RHE) while the E_{fb} of a p-type semiconductor must be positive of the oxygen evolution potential (1.23 V vs. RHE). The data presented in Fig. 6.4 show the illuminated OCP from a p-type GaPN grown with and without a Si p/n tandem junction. Because the E_{fb} of the non-tandem cell is negative (more cathodic) of the oxygen evolution potential, there is an energetic barrier to majority carrier hole injection at the counter electrode, and consequently an additional bias is required to overcome this barrier to drive the water oxidation reaction. Adding an integrated p/n-Si layer provides the necessary bias and shifts the OCP to values more positive (anodic) than the water oxidation potential.

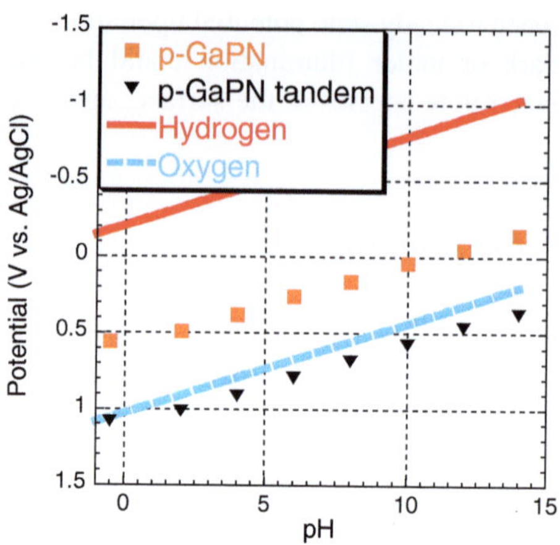

Fig. 6.4 E_{fb} versus pH for p-GaPN and for a p-GaPN tandem. Both materials show Nernstian behavior. The desired E_{fb} for n-type materials (not shown) should be above the solid line for the hydrogen evolution potential to enable majority electron carriers to drive the HER. Similarly, the E_{fb} for p-type materials (shown here) should be below the dashed line for the oxygen evolution potential to enable majority hole carriers to drive the OER

6.2 Mott–Schottky

6.2.1 Knowledge Gained

Determining E_{fb} is based on the Mott–Schottky (M–S) relationship involves measuring the capacitance of the space charge layer (C_{sc}) of the semiconductor electrode as a function of the applied potential (E) and applying the relationship according to Eq. (6.1) [9].

$$\frac{1}{C_{sc}^2} = \frac{2}{\varepsilon_r \varepsilon_0 A^2 e N_{Dopant}} \left(E - E_{fb} - \frac{kT}{e} \right) \tag{6.1}$$

where ε_r is the relative permittivity of the semiconductor (see Glossary for literature values for some semiconductor materials), ε_0 is permittivity in vacuum, A is the surface area, e is the charge of an electron, N_{Dopant} is the free carrier density, k is Boltzmann constant, T is the temperature, and E is the applied potential.

M–S is an electrochemical impedance spectroscopy (EIS) technique [10–12] that can be difficult to perform and interpret if the system is not ideal. When the measurement is successful, it is able to determine both the E_{fb} and the free charge carrier density (donors or acceptors, N_{Dopant}) of the photoelectrode. E_{fb}, along with the band gap (E_g) and the N_{dopant}, can be used to determine the band structure of a photoelectrode and if it possesses the proper alignment with respect to the water splitting potentials (see chapter "Introduction"). The N_{dopant} also plays a role in the bulk and surface semiconductor properties such as the width of the depletion layer and rate of recombination. The conductivity type is also revealed by M–S analysis. The M–S plot will possess a negative slope for p-type materials and a positive slope for n-type materials (positive slope). In the case that the M–S measurement is not successful, then other techniques such as Hall Effect can still yield conductivity and N_{dopant} for materials which can be deposited onto non-conductive substrates such as quartz.

6.2.2 Limits of Experiment

The primary disadvantage of M–S analysis is that the C_{sc} of a photoelectrode can be very complicated to measure depending on the sample composition and morphology. The ideal sample is a single-crystal material of high crystalline quality with moderate doping grown on an ohmic contact. Single-frequency measurements of the C_{sc} of an ideal sample are fairly straightforward. However, measurements for samples which are not ideal may require a full evaluation of the frequency range over many decades, followed by fitting of the data to an appropriate, but often complicated equivalent circuit model [13, 14].

Although the E_{fb} is ideally not dependent on the frequency used in an M–S measurement, the apparent value of C_{sc} can be complicated by contributions from surface state capacitance (C_{ss}) as well as double-layer capacitance (C_{dl}) at the semiconductor-electrolyte interface, giving a rise to a frequency dependence to the M–S results if the RC time constants for these elements are too similar. The frequency dispersion of capacitance indicates that the simple circuit model of a resistor and capacitor in series is inadequate and can result in unrealistic values for E_{fb} and N_{Dopant} which may not be representative of the true material properties. Typically, a frequency is chosen to be fast enough such that the timescale is too small to allow for effective filling and unfilling of surface states or for the buildup of a double-layer capacitance. The selection of frequency range is dependent on the crystallinity and other unique material properties of the sample. For example, a frequency range of 1–20 kHz may be applicable for highly crystalline materials. Researchers should consult with the literature for appropriate frequency ranges for their particular material.

M–S plots often exhibit frequency dependence. Gomes and Cardon [15, 16] divided the frequency dependent data into two general classes. In the first class, the M–S plots at different frequencies possess the same slope, but yield a different E_{fb} for each frequency (Fig. 6.5a). In the second class, the M–S plots have different slopes but converge to a common intercept, and hence yield a valid E_{fb} (Fig. 6.5b) but the interpretation of the C_{sc} is questionable and can lead to inaccurate values for carrier concentration. The behavior of this second class of data is attributed to the presence of surface irregularities and/or surface states in the semiconductor. It has been shown that the second type of M–S plot yields a valid value for the E_{fb}.

To accurately calculate N_{Dopant}, the actual surface area in contact with the electrolyte must be accurately determined. This surface area could be very different than that of the planar projected geometric area. In some cases, selective dye absorption to the surface can be used to quantify the roughness factor and obtain an electrochemically active surface area [18]. Other methods include evaluating the C_{dl} as a proxy for surface area using EIS measurements at low frequencies as well as cyclic voltammograms at various sweep rates [19].

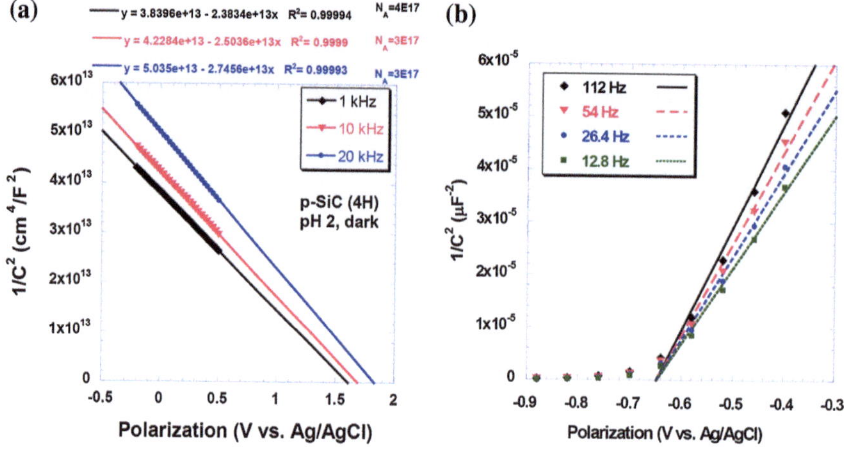

Fig. 6.5 (**a**) Mott–Schottky plots for single crystal p-SiC (the 4H polytype). The R^2 values near 1 point to an ideal response, but the frequency dispersion indicates the circuit model is not perfect (perhaps not accounting for surface state capacitance). Each data plot shows the forward and reverse scans and the lack of hysteresis indicates that the surfaces were not altered (damaged) during the data collection. Only the forward scanned data is displayed here. (**b**) Mott–Schottky plots at varying frequencies for Al-doped Fe_2O_3 [17]. Each of the four plots linearly extrapolates to the same E_{fb} of -0.65 V versus Ag/AgCl, but their slopes exhibit a frequency dependence. The calculated donor carrier density (N_D) is $1.35 \times 10^{23}/cm^3$ at 112 Hz, $1.51 \times 10^{23}/cm^3$ at 64 Hz, $1.67 \times 10^{23}/cm^3$ at 26.4 Hz, and $1.58 \times 10^{23}/cm^3$ at 12.8 Hz. These values for N_D are extremely high and indicate that the simple series RC circuit used to obtain C_{sc} is inadequate due to the presence of surface states or irregularities

6.2.3 Method

Refer to section "Cell Setup and Connections for Three- and Two-Electrode Configurations" for a three-electrode cell setup. The total preparation time is approximately 1 h and the experiment time requires anywhere from a few minutes to a few hours, depending on the number of potentials and frequencies used in the evaluation. The data analysis can be simple and straightforward for ideal samples, taking only a few minutes. Non-ideal M–S plots, however, can take much longer to analyze, and may require the researcher to move on to another technique to analyze E_{fb}.

6.2.3.1 Equipment

- Potentiostat with a frequency response analyzer (FRA) or with EIS capabilities.
- Three-electrode electrochemical cell: WE (Semiconductor material), CE, and RE (e.g., Ag/AgCl in saturated KCl).
- Electrolyte solutions:

- It is advisable to perform the M–S test with 2–3 different pH solutions to assess the pH dependence of the E_{fb}. The pH of the solution and the electrolyte will be dependent on semiconductor being evaluated (see Table 3.1 in section "Cell Setup and Connections for Three- and Two-Electrode Configurations").
- Solutions should be sparged with the gaseous product of the photoelectrode (H_2 for p-type photocathodes and O_2 for n-type photoanodes) to produce a well-defined redox potential and consistent Schottky barrier height.

High quality samples of semiconductors such as $GaInP_2$, GaN, or SiC can be employed as positive controls. These materials produce close to ideal M–S plots and thus tend to yield more consistent E_{fb} and N_{dopant} information.

6.2.3.2 Experimental Procedure

The M–S experiment should be first carried out in the dark at a single frequency over a range of potentials to determine the voltage dependence of the space charge capacitance. This is then repeated for five or more different frequencies in the frequency range that is appropriate for the material tested. Starting at a small amplitude of 5–10 mV minimizes non-linear effects from larger amplitudes, although larger amplitudes may need to be employed if the signal is too low (which manifests as noisy, scattered data points). Take care to avoid frequencies of electrical equipment such as 60 Hz in the U.S.A. or multiples of these frequencies. Note that the experiment can also be carried out under illumination, but this approach will not be discussed here.

The experiment should be carried at reverse bias potentials relative to the expected E_{fb} (which can be initially determined from an illuminated OCP measurement). For n-type (p-type) photoanodes (photocathodes), the scan should be performed at potentials anodic (cathodic) of OCP to a couple hundred millivolts before approaching OCP and back to the reverse bias potential. In addition, the potentials should be chosen to avoid any Faradaic reactions in order to prevent further complication from a charge transfer resistance.

If good results (linear M–S plots) are obtained with the first pH solution, then testing should be repeated with several different pH solutions to evaluate the pH dependency of the E_{fb}.

The electrical wiring plays an important role in impedance measurements in general. Care should be taken to minimize the cable length and to employ proper shielding (co-axial cable) to minimize effects from inductance or parasitic series resistance and capacitance.

6.2.4 Data Analysis

The E_{fb} determined from capacitance measurements is based on the well-known Mott–Schottky relation shown in Eq. (6.1). This analysis is accurate when the capacitance of the space charge layer is well distinguished from the capacitance of other elements in the material such as due to charging at surface states or buildup of the double-layer, which may not necessarily be the case. Because these other capacitances are in series with the capacitance of the space charge region, the smallest capacitance (typically the space charge region) can often dominate the response of the overall capacitance [20], although this may not always be the case.

For an ideal photoelectrode, the equivalent circuit can be simplified to a resistor (R) and a capacitor (C) in series. The R represents the resistance of the semiconductor bulk (plus any series resistance from the electrode wires and the electrolyte), and the C represents the capacitance of the space charge region (C_{sc}). For an ideal system, a plot of $1/C_{sc}^2$ versus electrode potential (E) yields a straight line. The line is extrapolated to $1/C_{sc}^2 = 0$. The x-intercept equals $E_{fb} + kT/e$ and the slope is proportional to the charge carrier concentration or doping density (N_{Dopant}), as shown by Eq. (6.2). This equation is obtained by substituting the relevant terms from Eq. (6.1).

$$N_D(\mathrm{cm}^{-3}) = \frac{1.41 \times 10^{32}(\mathrm{cm} \times F^{-2} \times V^{-1})}{\varepsilon_r \times A^2(\mathrm{cm}^4) \times \mathrm{slope}(F^{-2} \times V^{-1})} \tag{6.2}$$

The relative permittivity of the semiconductor (ε_r) is often assumed to be 10 because it is usually slightly above or below this value [21, 22]. However, this is dependent on each material. Researchers should consult the literature for values appropriate for their materials. Typical N_{Dopant} values for semiconductors are in the range of 10^{15}–10^{18} carriers/cm^3. Calculated N_{Dopant} values outside of this range may be questionable. If the surface area of the electrode is not correct, then the doping density will also not be correct. If the space charge is populated with electrons, then the slope is negative, resulting in a negative doping density and an indication that the semiconductor is a p-type material. The opposite is true for n-type materials, which will exhibit a positive slope.

If the M–S plot is not linear over the entire measured potential range, then one can attempt to fit only the linear portion (200 mV range minimum) of the plot to a line. However, accuracy will likely be poor due to the subjectivity in this approach. Some plots may never exhibit a linear portion, as shown in Fig. 6.6.

The M–S measurement is performed at several different frequencies to verify that the slope and x-intercept do not change. If they do change and/or the M–S plot is not linear, as is the case in Fig. 6.6, then a more detailed analysis using frequencies over multiple decades (e.g. 1×10^6–1×10^{-2} Hz) should be used in conjunction with a more sophisticated equivalent circuit model [14, 23].

Once the E_{fb} for each pH is determined, plot the E_{fb} as a function of pH in order to determine the dependence of the band structure on pH as explained previously in section "Open-Circuit Potential and pH".

Fig. 6.6 Non-ideal M–S plots for n-SiC (4H polytype). The absence of a linear region as well as the presence of hysteresis indicates a change in the surface during the measurements, possibly due to corrosion. Fitting the data yields inconsistent E_{fb} and N_{dopant} values at several frequencies

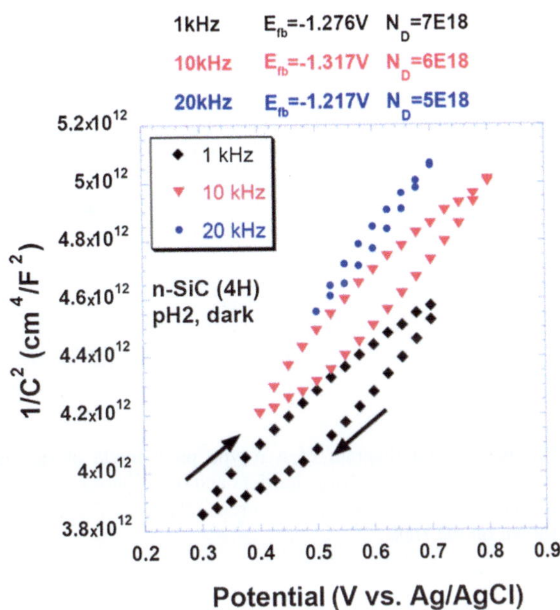

6.3 Three-Electrode j–V and Photocurrent Onset

Research in photoelectrochemistry (PEC) is mainly focused on the discovery of semiconductor materials that are able to dissociate water using only photogenerated charge carriers without an external bias. This capability can be evaluated by shorting the PEC electrode directly to a counter electrode and measuring the photocurrent generated under AM 1.5 G illumination (see chapter "Two-Electrode Short Circuit and j–V"). However, from a materials research and development point of view, it is important to evaluate the fundamental properties of a photoelectrode, such as its potential for photocurrent onset as well as its ability to generate photocurrent at a given potential. These properties can be determined from a j–V analysis of a material in a three-electrode configuration under illumination. In addition, this technique can be used to estimate the material flat-band potential (E_{fb}).

6.3.1 Potential Range of Photocurrent Generation

Prior to the j–V analysis, the conductivity type (p or n) of the semiconductor should be known. The experiment should be carried out under small forward bias potentials (i.e. at potentials between the flat-band potential and the open circuit potential) towards reverse bias potentials. It is important to keep in mind that

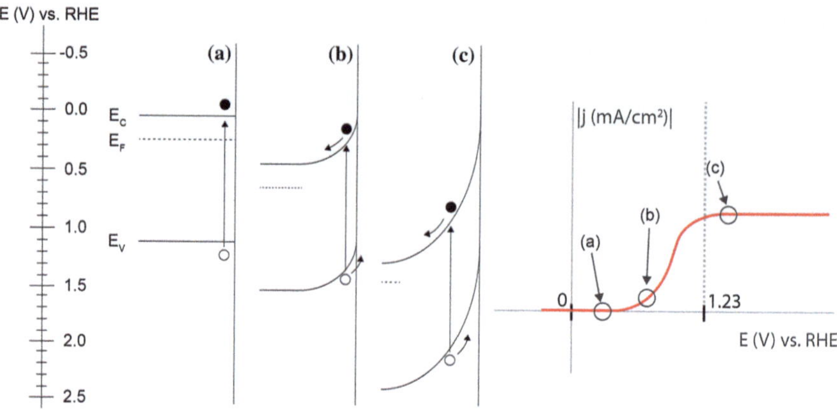

Fig. 6.7 Band diagram of a n-type photoanode at (**a**) flat-band potential, (**b**) a potential sufficient to separate charge carriers and drive photocurrent, and (**c**) large reverse bias potential sufficient to saturate the photocurrent response. The corresponding hypothetical j–V *curve* is shown on the *right*

forward biasing most p-type electrodes to potentials more anodic than their E_{fb} can often lead to their degradation through anodic stripping and should be avoided. Similarly, forward biasing n-type electrodes to potentials more cathodic than their E_{fb} can lead to cathodic corrosion reactions.

When illuminated with energy equal to or above the band gap ($hv \geq E_g$) at these operating potentials, minority hole carriers in n-type electrodes drive the OER at the electrode–electrolyte interface while minority electron carriers in p-type electrodes drive the HER at this interface. The potential at which this phenomenon begins to occur is the photocurrent onset potential (E_{onset}), which is offset relative to the flat-band potential (E_{fb}) by the required kinetic overpotentials for the reaction of interest. The difference between the photocurrent onset potential (E_{onset}) and the reversible redox potential of interest (E^0) is the onset voltage (V_{onset}). A band diagram of a n-type photoanode and its hypothetical j–V response is shown in Fig. 6.7.

At higher reverse bias, a dark current will eventually appear due to lowering of the Schottky barrier height, reverse bias breakdown of the semiconductor, or shunting of majority charge carriers through conductive defect states in the electrode (effectively negating the Schottky barrier). A more detailed discussion can be found elsewhere [24, 25]. In this region the surface of the semiconductor behaves as a metal and provides additional dark current in addition to the photocurrent. The potential $E_{dark\text{-}onset}$ at which dark current becomes significant can be determined by sweeping the potential from E_{onset} in a three-electrode configuration toward higher reverse bias conditions without illumination (see Fig. 6.7). It follows that j–V characterization in a three-electrode configuration should be performed between $E_{dark\text{-}onset}$ and E_{onset}. The same potential range can be used for subsequent PEC experiments such as photocurrent spectroscopy. This range may be pH dependent, as the band edge potentials follow a Nernstian dependence on [H$^+$] for many semiconductor electrodes.

6.3.2 Knowledge Gained

In order to achieve water splitting, some PEC devices require a potential bias. In some standalone configurations, such as the tandem hybrid device, such a bias can be provided internally by a photovoltaic junction stacked in series with the PEC junction. The determination of the saturated photocurrent density achievable by a PEC material, as well as the bias required to achieve this value, is crucial in identifying possible integration schemes for the material of interest.

The photocurrent onset can be used to estimate the E_{fb}, and can be determined by measuring the photocurrent density (j_{ph}) as a function of the potential versus a reference electrode (E_{ref}) and then comparing the dark and illuminated scans. As the potential moves towards greater reverse bias, the electric field generated across the space charge layer increases and drives charge separation and charge transfer to the electrolyte, thereby enhancing the photocurrent as shown in Fig. 6.7. Under ideal circumstances, the potential at which the semiconductor changes from accumulation to depletion coincides with the onset of photocurrent and this potential is taken as the E_{fb}, although this is often offset by the required kinetic overpotential to drive the reaction.

To illustrate the metrics that should be used to assess device performance in a three-electrode measurement, Fig. 6.8a, b show ideal illuminated cyclic voltammograms (CV) for a hypothetical photoanode and photocathode, respectively, on a reversible hydrogen electrode (RHE) scale, where 0 V represents the reversible potential for the hydrogen evolution reaction (HER) and 1.23 V represents the reversible potential for the oxygen evolution reaction (OER). Several key metrics are labeled, including the current density at the reversible potential for HER ($j^o_{E=HER}$), the current density at the reversible potential for OER ($j^o_{E=OER}$), and the underpotential for photocurrent onset relative to OER or HER (V$_{onset, OER}$ and V$_{onset, HER}$) which are measures of a voltage difference between two potentials (e.g. V$_{onset,OER}$ = E$^o_{OER}$ − E$_{onset,OER}$). In addition, a fill factor (ff) analogous to the same metric used in the PV community describes how quickly the photocurrent rises versus potential. In order to maximize performance, |V$_{onset}$|, $j^o_{E=HER}$, $j^o_{E=OER}$, and ff should be as high as possible. However, it is important to note that unlike a PV device, these terms cannot be conveniently combined to yield a valid power conversion efficiency, since |V$_{onset}$| is measured on a three-electrode scale and only encapsulates the voltage between the WE and RE, and not the full cell voltage between the WE and CE (see Chapter "Efficiency Definitions in the Field of PEC"). For this discussion, V_{onset} is not referred to as V_{oc} (open circuit voltage) in order to avoid encouraging the calculation of a power conversion efficiency.

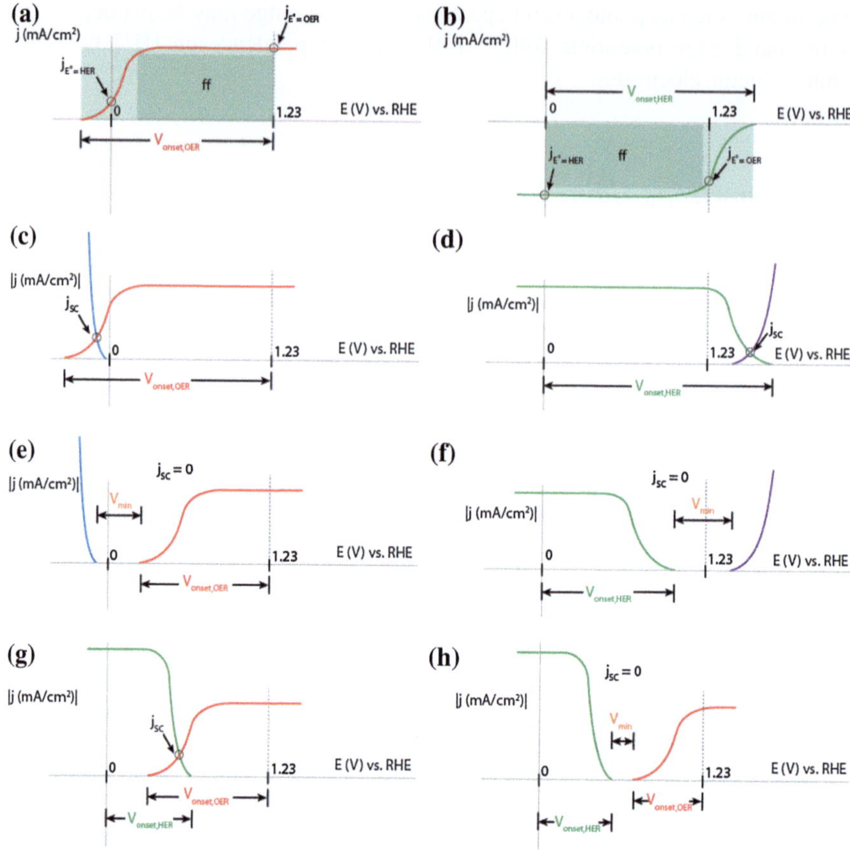

Fig. 6.8 Idealized illuminated cyclic voltammograms for (**a**) photoanode, (**b**) photocathode, (**c**) photoanode with large $V_{onset,OER}$ overlayed with dark current from a HER cathode, (**d**) photocathode with large $V_{onset,HER}$ overlayed with dark current from a OER anode, (**e**) photoanode with small $V_{onset,\,OER}$ overlayed with dark current from a HER cathode, (**f**) photocathode with small $V_{onset,\,HER}$ overlayed with dark current from an OER anode, (**g**) photoanode overlayed with a photocathode in which the combination of V_{onset} sufficiently overlaps to drive spontaneous water splitting, and (**f**) photoanode overlayed with a photocathode in which their combined V_{onset} values are insufficient to overlap. *Red* = photoanode current under illumination, *purple* = dark current for OER anode, *green* = photocathode current under illumination, *blue* = dark current for HER cathode. In (**c**–**f**), the cathode/photocathode current is flipped positive for analysis of j_{sc} with photoanode. V_{min} represents the minimum applied potential required to drive water splitting in cases where $j_{SC} = 0$

It is possible to extract a maximum theoretical j_{SC} (and consequently STH) from three-electrode measurements, but only if at least TWO measurements are made for both the cathode and anode. Figure 6.8c–h depicts several typical scenarios. In Fig. 6.8c, the photoresponse of a photoanode is overlayed with the dark current response of a cathode catalyst (note that the sign of the current response for the cathode is flipped for this analysis). Since a complete device under operation drives the same current through the anode and cathode to maintain total charge neutrality, the intersection between the two curves represents the theoretical j_{SC} value. Figure 6.8d shows a similar configuration, except with a photocathode and anode catalyst. Figure 6.8e, f are cases for a photoanode and photocathode, respectively, in which the semiconductor cannot generate a sufficient V_{onset} to overlap with the respective counterelectrode. As such, the j_{SC} is zero for these cases and there is no ability to spontaneously drive water splitting. Figure 6.8g, h represent a tandem cell in which a photocathode and photoanode are stacked in series. In Fig. 6.8g, the intersection between the two photoresponse curves equals the theoretical j_{SC} of the device, whereas Fig. 6.8f shows a tandem cell in which the voltage generated by both components is still insufficient to split water. In this case, j_{SC} is once again zero. For simplicity, dark current is not shown for the photoelectrodes in Fig. 6.8.

6.3.3 Limits of Experiment

6.3.3.1 Confidence in Results

In this experiment, potentials are reported versus a reference electrode. Since several different reference electrodes may be used in a lab and their potential may drift with time, it is important to measure the potential and verify that each individual RE potential is stable in order to compare results. Details on RE stability evaluation are presented in section "Cell Setup and Connections for Three- and Two-Electrode Configurations".

The E_{fb} is a property of the semiconductor interface that depends on the electrolyte in which the measurement is made. The onset of photocurrent does not necessarily define the E_{fb} potential because other interfacial effects may delay the onset to a point beyond the transition from accumulation to depletion. The error from such interfacial effects could be on the order of a few millivolts to over a volt. One such interfacial effect might be the kinetic overpotential required to drive the reaction. This overpotential shifts the photocurrent onset in the cathodic direction for p type samples and in the anodic direction for n type samples. Therefore, catalysts are often deposited onto the electrode surface to minimize the overpotentials (see section "Catalyst Surface Treatments"). However, the modification of electrode surfaces with catalysts may influence the semiconductor/electrolyte junction and surface states and additionally shift the E_{fb} in unexpected ways. Ideally, the catalyst treatment will improve the accuracy of the E_{fb} measured by this technique, although effects such as Fermi level pinning may introduce a change in the band structure at the surface which may negate the improvement from a reduced kinetic overpotential.

6.3.3.2 Source of Errors

Accurate surface area determination is important for reporting current density. One may also verify that the RE chosen for the analysis is compatible with the electrolyte. More details on counter electrode selection as well as RE compatibility with electrolytes are presented in section "Cell Setup and Connections for Three- and Two-Electrode Configurations". Dark scan current should be subtracted from j–V characteristics under illumination if one wants to determine the E_{fb}. Fast scan rates (higher than 100 mV/s) will give rise to large capacitive currents and should be avoided, particularly for high surface area electrodes. If corrosion is occurring, either due to chemical (e.g. oxidation) or mechanical (vigorous bubbling) processes, there may be error in the E_{fb} because of changes in the surface composition or morphology. Thus, careful observation and suitable physical (optical microscopy) and chemical analysis of the material surface before and after the test may be necessary to determine the presence of corrosion. Depending on the electrode preparation technique (see section "Electrode Preparation"), epoxy might be required to cover electrical leads soldered to the substrate. However, incompatibility between the epoxy and the electrolyte may expose the metal contacts and convolute the measurement. In addition, the chemical components of the degraded epoxy may adsorb to the semiconductor surface and shift the E_{fb}. Finally, the solar simulator should be calibrated for AM 1.5 G illumination in order to obtain representative current densities for solar illumination.

6.3.3.3 Pitfalls of Experiment

An important limitation is that the three-electrode j–V measurement cannot be utilized to calculate a power conversion efficiency because the three-electrode scale represents only the half-cell voltage between the WE and RE (see Chapter "Efficiency Definitions in the Field of PEC"). To emphasize why a "half-cell" efficiency is insufficient, Fig. 6.9a illustrates the case for two distinctly different hypothetical photoanodes. Photoanode A possesses a small $V_{onset,\ OER}$ and large $j^o_{E=\ OER}$ while Photoanode B possesses a large $V_{onset,OER}$ and small $j^o_{E=\ OER}$. Both photoanodes in this case would seemingly yield the same half-cell efficiency described by in Eq. (2.10) in Chapter "Efficiency Definitions in the Field of PEC" using the "operating" points V_{op} and j_{op} that correspond to the point of maximum fill factor. However, further analysis in Fig. 6.9b reveals that a complete device which includes the addition of a photocathode would clearly distinguish Photoanode B as the superior component.

This approach to assessing STH from two overlayed three-electrode measurements often overestimates the true two-electrode performance, mainly because integration of multiple material components is rarely a perfect process. Factors include lost illumination (e.g. catalyst shadowing, mismatched refractive indices resulting in increased reflection, absorption mismatched band gaps), ohmic resistance (iR) losses, and non-ideal interfacial engineering of the various

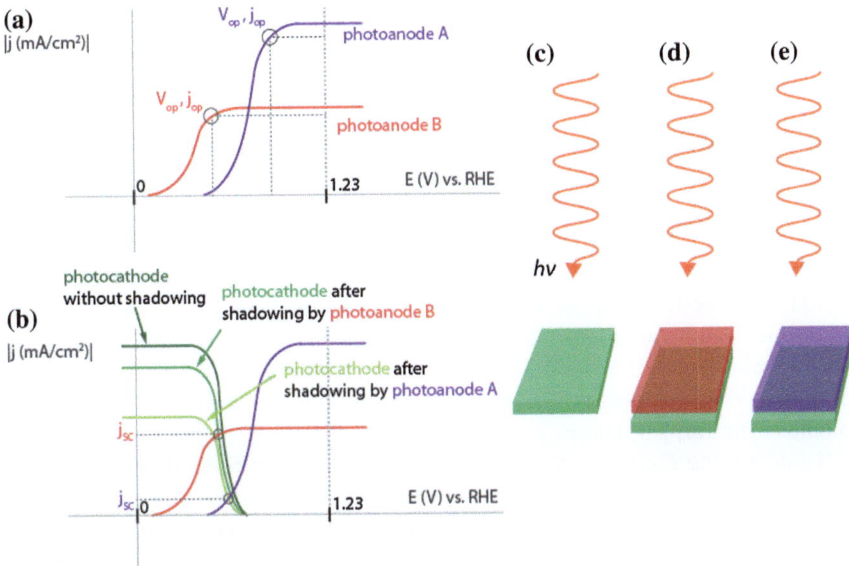

Fig. 6.9   The case of two photoanodes (*A* and *B*) with distinctly different j–V responses that would yield similar efficiencies if analyzed as half-cells (**a**) whereas correct evaluation in a full device (**b**) would clearly distinguish B as the better component that leads to a greater j_{SC}. Also shown are the effects of shadowing on the photocathode. In comparison to a standalone photocathode (**c**), the photocurrent produced by a photocathode placed in tandem with photoanode B (**d**) will be greater than if it were placed in tandem with photoanode A (**e**)

components resulting in increased recombination. Using data generated from two different illumination sources can also be a major source of error. The effect of absorption shadowing is shown using the hypothetical photoelectrodes in Fig. 6.9b. When the photocathode is placed under the photoanode, the photoanode will absorb a fraction of the illumination above its band gap and pass lower energy photons to the photocathode, thereby lowering the photocurrent generated in the photocathode. While the approach overlaying three-electrode plots allows one to rapidly assess the potential performance for a photoelectrode, it must be noted that this produces a theoretical maximum j_{SC} value, and will always overestimate the performance of an integrated device. As such, the best way to evaluate STH is still with a two-electrode measurement (see chapter "Two-Electrode Short Circuit and j–V").

6.3.4   Method

The following procedures involve performing PEC characterizations in a three-electrode configuration using a reference electrode. A complete description of usable test cell setups, reference electrodes, and electrolytes is presented in section "Cell Setup and Connections for Three- and Two-Electrode Configurations".

6.3.4.1 Determining the Potential Range of Photocurrent Generation

As discussed in section "Potential Range of Photocurrent Generation", the conductivity type of the semiconductor determines the direction of the potential scan (anodic of E_{onset} for n-type and cathodic of E_{onset} for p-type). The initial potential should be close to E_{onset}. In theory, E_{onset} should be close to E_{fb}, which can be determined using an illuminated OCP measurement (see section "Illuminated Open-Circuit Potential (OCP)"). In practice these values can differ by a few hundred mV. Consequently, the user may need to adjust the window of the sweep to collect the full j–V curve.

To determine $E_{dark\text{-}onset}$, an initial sweep should be performed without illumination. This sweep should be interrupted when dark current becomes significantly large. A second sweep is performed to determine the photocurrent onset potential under illumination between $E_{dark\text{-}onset}$ and the illuminated OCP. The potential at which the direction of the photocurrent changes (i.e. from anodic to cathodic for n-type materials) is the photocurrent onset potential.

6.3.4.2 Scan Rate and Light Chopping

The scan rate must be adapted to account for variations in material characteristics. High surface area electrodes will produce a much higher double-layer capacitance and will require slower scan rates in order to minimize the error introduced by this capacitance current. In addition, both photocurrent and dark current measurements can be performed at the same time by chopping the light in order to limit the number of sweeps and the duration that the electrode is immersed in the electrolyte to minimize corrosion. This can be performed either manually or by using a wheel or a controlled shutter. The chopping frequency generally ranges from 0.5 Hz to 0.05 Hz, depending on the scan rate, and is selected in order to minimize errors from transient photocurrent behavior which occurs due to capacitance charging upon turning the light on and off.

6.3.4.3 Photocurrent Onset Measurement to Estimate Flat-Band Potential

To determine the E_{fb} with this technique, both the dark current and the photocurrent need to be measured. If dark current is not negligible over the potential range of interest, then it should be subtracted from the total measured current under illumination to produce a *j–V* curve which only contains the photocurrent. This analysis can be repeated in different pH solutions to verify the Nerstian dependence of E_{fb} with pH. To limit the effect of kinetic overpotentials on E_{fb} estimation, this analysis can be performed on the same PEC-electrode after a catalytic surface treatment, or using more reversible redox couples such as $Fe(CN)_6^{4-/3-}$. These redox reactions are kinetically facile and typically do not require catalyst treatments on the surface. Due to the smaller overpotentials, the estimated onset potential can be assumed to be closer to the true E_{fb} value for the semiconductor-electrolyte interface.

6.3.5 *Time Required for Preparation, Experiment, and Data Treatment*

Depending on the light source used for the experiment, several tens of minutes might be required to achieve steady power illumination and stable spectral distribution. In the meantime, the WE, CE, and RE can be placed in the electrochemical cell and the electrolyte can be prepared and sparged with an inert gas such as Ar or N_2. The typical experiment duration is a few minutes to tens of minutes per device, depending on the sweep rate, potential range, and number of sweeps. Data processing consists of normalizing the photocurrent to the illuminated area as well as making plots of j_{ph} [25] versus V to determine photocurrent onset [26].

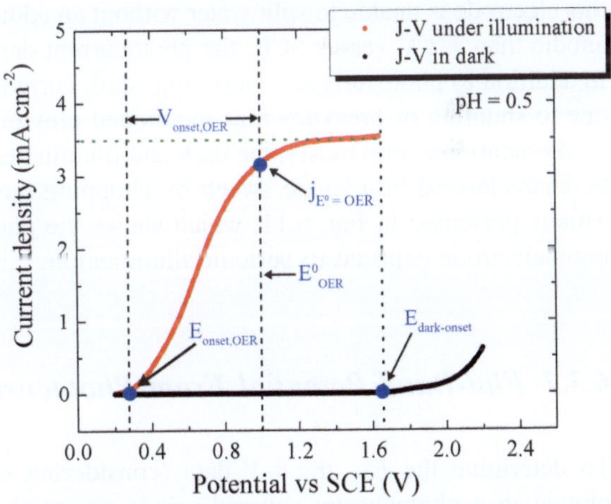

Fig. 6.10 Three-electrode *j–V* plot of a WO$_3$-based PEC electrode in dark (*black curve*) and under AM 1.5 G illumination (*red curve*). Measurement was performed in a 0.33 M H$_3$PO$_4$ electrolyte using a scan rate of 25 mV/s

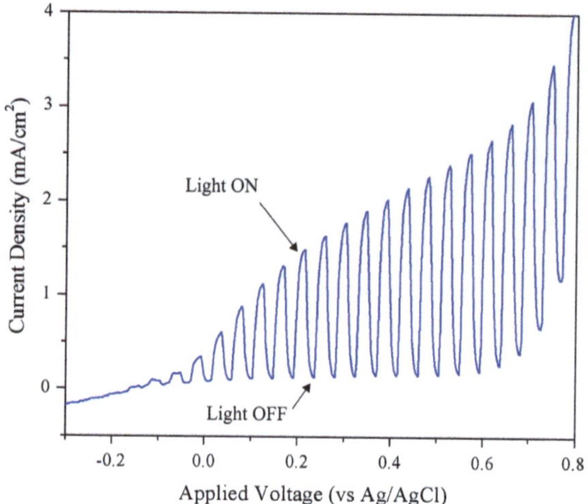

Fig. 6.11 j–V characteristic of a Fe$_2$O$_3$-based PEC electrode in 0.5 M KNO$_3$ at an illumination intensity of 400 mW/cm^2. The light was chopped at a frequency of 0.5 Hz using a scanning rate of 100 mV/s

6.3.6 Data Analysis and Expected Results

6.3.6.1 Potential Range and Saturated Photocurrent

The result of this analysis is a plot of photocurrent density as a function of the potential measured versus a reference electrode. An example j–V curve for a WO$_3$ film is shown in Fig. 6.10. In this case, the photocurrent onset occurs at approximately 0.27 V versus SCE (0.54 V vs. RHE) which corresponds to a $V_{\text{onset,OER}}$ of 0.69 V (1.23–0.54 V). Since the onset of photocurrent does not occur cathodic of the reversible potential for HER ($E_{\text{HER}}^0 = -0.27$ V versus SCE in this electrolyte), this electrode is unable to split water without an additional bias. At potentials more anodic than 1.2 V versus SCE, the photocurrent density saturates at 3.5 mA/cm^2. In addition to photocurrent, reverse bias dark current onsets at 1.65 V versus SCE due to shunting or breakdown as mentioned previously.

As mentioned previously, the dark and illuminated behavior of an electrode can be characterized in a single sweep by chopping the light at a regular frequency. This is presented in Fig. 6.11, which shows the photoresponse of a Fe$_2$O$_3$-based photoelectrode exposed to periodic illumination.

6.3.7 Flat-Band Potential From Photocurrent Onset

To determine the E_{fb}, the j–V data (considering only the photocurrent, j_{ph}) is plotted in a photocurrent squared versus potential (j_{ph}^2 vs. V) [26]. The linear portion closest to the onset of photocurrent is fitted to a linear line. The x-intercept

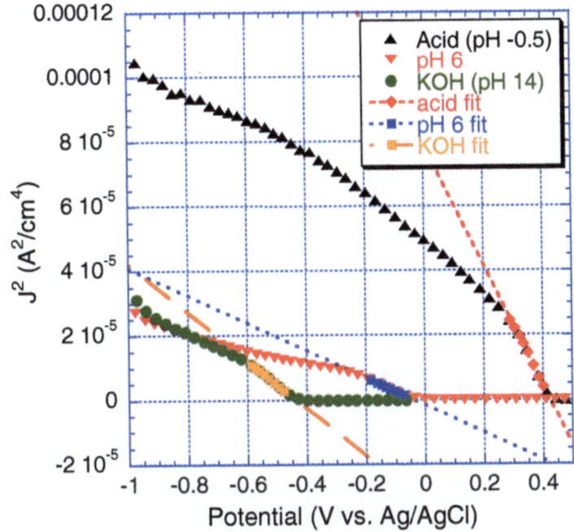

Fig. 6.12   Photocurrent density squared (J^2) versus potential plot and their linear fits used to find the onset of photocurrent in three different pH electrolyte solutions. The x-intercepts are the flat-band potentials for each pH: 0.4 V for pH –0.5, –0.05 V for pH 6, and –0.45 V for pH 14. This data was taken on a Pt-catalyzed p-GaPN photoelectrode that exhibited cathodic photocurrent [27]. The approximately 59 mV/pH shift indicates a Nernstian dependence of the surface with pH

of the fitted line is considered to be the E_{fb} (shifted by any necessary overpotential to drive the redox reaction), as presented in Fig. 6.12. The actual E_{fb} should be more cathodic for photoanodes and more anodic for photocathodes by a value approximately equal to the kinetic overpotential required to drive each respective reaction.

The dependence of E_{fb} as a function of pH can be assessed, and if the pH dependence is −59 mV/pH, then it is a Nernstian dependence, which usually indicates that the surface of the semiconductor may be protonated or hydroxylated depending on the concentration of H^+ and OH^- [28]. It is useful to compare the pH dependence of E_{fb} with the pH dependence of the redox potential of hydrogen and oxygen evolution to determine if the band edges of the semiconductor are aligned to photoelectrochemically split water at any pH.

The E_{fb} can also be determined using redox couples such as $Fe(CN)_6^{4-/3-}$ instead of the hydrogen or oxygen evolution reactions. The choice of redox couple depends upon the relative position of its equilibrium potential with respect to the band edge positions of the semiconductor. It is also desirable to conduct photo-current onset potential measurements using several redox couples in both acidic and basic electrolytes. The validity of the data needs to be considered by comparing data obtained using different redox couples. The interactions of surface states with redox species in solution can shift the potentials. If the data sets are not consistent, then further investigation is required.

References

1. A.J. Nozik, R. Memming, Physical chemistry of semiconductor–liquid interfaces. J. Phys. Chem. **100**, 13061–13078 (1996)
2. T.G. Deutsch, C.A. Koval, J.A. Turner, III–V nitride epilayers for photoelectrochemical water splitting: GaPN and GaAsPN. J. Phys. Chem. B **110**, 25297–25307 (2006)
3. A.J. Bard, A.B. Bocarsly, F.R.F. Fan, E.G. Walton, M.S. Wrighton, The concept of fermi level pinning at semiconductor–liquid junctions—consequences for energy-conversion efficiency and selection of useful solution redox couples in solar devices. J. Am. Chem. Soc. **102**, 3671–3677 (1980)
4. A.J. Bard, F.-R.F. Fan, A.S. Gioda, G. Nagasubramanian, H.S. White, On the role of surface states in semiconductor electrode photoelectrochemical cells. Faraday Discuss. Chem. Soc. **70**, 19–31 (1980)
5. J.F. Geisz, D.J. Friedman, S. Kurtz, GaNPAs solar cells lattice-matched to GaP. Photovoltaic Specialists Conference 2002. Conference Record of the Twenty-Ninth IEEE (2002)
6. T.G. Deutsch, Sunlight, Water, and III-V Nitrides for Fueling the Future. Ph.D. Thesis, University of Colorado (2006)
7. Z. Chen, T.F. Jaramillo, T.G. Deutsch, A. Kleiman-Shwarsctein, A.J. Forman, N. Gaillard, R. Garland, K. Takanabe, C. Heske, M. Sunkara, E.W. McFarland, K. Domen, E.L. Miller, J.A. Turner, H.N. Dinh, Accelerating materials development for photoelectrochemical hydrogen production: standards for methods, definitions, and reporting protocols. J. Mater. Res. **25**, 3–16 (2010)
8. Y.V.P. Viktor, A. Myamlin, *Electrochemistry of Semiconductors* (Plenum Press, New York, 1967)
9. M.E. Orazem, B. Tribollet, *Electrochemical Impedance Spectroscopy* (*The ECS Series of Texts and Monographs*) (Wiley, New York, 2008)
10. A.J. Bard, L.R. Faulkner, *Electrochemical Methods: Fundamentals and Applications* (Wiley, New York, 2001)
11. M.D. Archer, A.J. Nozik, *Nanostructured and Photoelectrochemical Systems for Solar Photon Conversion*, vol 3 (Imperial College Press, London, 2008)
12. R. Memming, *Semiconductor Electrochemistry* (Wiley–VCH, Weinheim, 2001)
13. E. Barsoukov, J.R. Macdonald, *Impedance Spectroscopy: Theory, Experiment, and Applications* (Wiley, New York, 2005)
14. F.L. Formal, N. Tétreault, M. Cornuz, T. Moehl, M. Gratzel, K. Sivula, Passivating surface states on water splitting hematite photoanodes with alumina overlayers. Chem. Sci. **2**, 737–743 (2011)
15. E.C. Dutoit, RLv Meirhaeghe, F. Cardon, W.P. Gomes, Investigation on the frequency-dependence of the impedance of the nearly ideally polarizable semiconductor electrodes CdSe, CdS and TiO_2. Berichte der Bunsengellschaft fur physikalische Chemie **79**, 1206–1213 (1975)
16. A.J. Nozik, Photoelectrochemistry: Applications to solar energy conversion. Ann. Rev. Phys. Chem. **29**, 189–222 (1978)
17. Z. Chen, T.F. Jaramillo, T.G. Deutsch, A. Kleiman-Shwarsctein, A.J. Forman, N. Gaillard, R. Garland, K. Takanabe, C. Heske, M. Sunkara, E.W. McFarland, K. Domen, E.L. Miller, J.A. Turner, H.N. Dinh, Accelerating materials development for photoelectrochemical hydrogen production: Standards for methods, definitions, and reporting protocols. J. Mater. Res. **25**, 3–16 (2010)
18. A. Kay, I. Cesar, M. Gratzel, New benchmark for water photooxidation by nanostructured alpha-Fe_2O_3 films. J. Am. Chem. Soc. **128**, 15714–15721 (2006)
19. J.D. Benck, Z. Chen, L.Y. Kuritzky, A.J. Forman, T.F. Jaramillo, Amorphous molybdenum sulfide catalysts for electrochemical hydrogen production: insights into the origin of their catalytic activity (ACS Catalysis, 2012), pp. 1916–1923

20. J.A. Turner, Energetics of the semiconductor-electrolyte interface. J. Chem. Educ. **60**, 327–329 (1983)
21. D.E. Aspnes, A.A. Studna, Dielectric functions and optical parameters of Si, Ge, GaP, GaAs, GaSb, InP, InAs, and InSb from 1.5 to 6.0 eV. Phys. Rev. B **27**, 985–1009 (1983)
22. K.B. Kahen, J.P. Leburton, General theory of the transverse dielectric constant of III-V semiconducting compounds. Phys. Rev. B **32**, 5177–5184 (1985)
23. L. Badia-Bou, E. Mas-Marza, P. Rodenas, E.M. Barea, F. Fabregat-Santiago, S. Gimenez, E. Peris, J. Bisquert, Water oxidation at hematite photoelectrodes with an iridium-based catalyst. J. Phys. Chem. C **117**, 3826–3833 (2013)
24. R.F. Pierret, *Semiconductor Device Fundamentals*. pp. 490-492 (McGraw-Hill, New York, 1996)
25. F.C.W.P. Gomes, W. Dekeyser, *Photovoltaic and Photoelectrochemical Solar Energy Conversion* (Plenum Press, New York, 1981)
26. D.S. Ginley, M.A. Butler, Photoelectrolysis of water using iron titanate anodes. J. Appl. Phys. **48**, 2019–2021 (1977)
27. T.G. Deutsch, Sunlight, Water, and III-V Nitrides for Fueling the Future. Ph.D. Thesis, University of Colorado (2006)
28. J.A. Turner, Energetics of the semiconductor–electrolyte interface. J. Chem. Educ. **60**, 327–329 (1983)

Chapter 7
Incident Photon-to-Current Efficiency and Photocurrent Spectroscopy

7.1 Knowledge Gained

The incident photon-to-current efficiency (IPCE) is a measure of the ratio of the photocurrent (converted to an electron transfer rate) versus the rate of incident photons (converted from the calibrated power of a light source) as a function of wavelength. IPCE takes into consideration the efficiencies for photon absorption/charge excitation and separation $\left(\eta_{e^-/h^+}\right)$, charge transport within the solid *to* the solid–liquid interface $(\eta_{transport})$, and interfacial charge transfer *across* the solid–liquid interface $(\eta_{interface})$ as per Eq. 2.4 in Chapter "Efficiency Definitions in the Field of PEC".

$$IPCE = EQE = \eta_{e^-/h^+}\,\eta_{transport}\,\eta_{interface}$$

The IPCE measurement gives *combined* information about these efficiencies; however, experimental strategies exist that allow each component to be separated by employing methods that identify the dominating (hindering) efficiency factor. For example, if the η_{e^-/h^+} (by employing a UV–Vis measurement as per Chapter "UV-Vis Spectroscopy") and IPCE of the sample are known, $\eta_{transport}$ and $\eta_{interface}$ at the specific test conditions can be deconvoluted. Consequently, if a material has a low IPCE and a high η_{e^-/h^+}, then one may conclude that $\eta_{transport}$ and/or $\eta_{interface}$ are limiting. To help decouple these last two effects, an electrocatalyst can be deposited on the surface to improve the kinetics of the reaction in order to enhance $\eta_{interface}$ and observe the effect upon IPCE. If no enhancement is observed, then $\eta_{transport}$ is the limiting factor. Ideally, addition of an electrocatalyst would only change $\eta_{interface}$, but it may also change $\eta_{transport}$ if one considers carrier transport across the newly created semiconductor—electrocatalyst interface. It is also possible that the electrocatalyst, or the procedure used for its deposition, may modify the semiconductor in unintended beneficial or detrimental ways, such as removal or introduction of surface trap states. Detailed discussion of these phenomena is outside the scope of this document.

Z. Chen et al., *Photoelectrochemical Water Splitting*,
SpringerBriefs in Energy, DOI: 10.1007/978-1-4614-8298-7_7,
© The Author(s) 2013

Measuring the IPCE is also useful to determine the band gap. The band gap derived from IPCE may be higher than that obtained by optical spectroscopy techniques (see Chapter "UV-Vis Spectroscopy"), since the onset of photocurrent may be limited by slow kinetics $\left(\eta_{\text{interface}}\right)$ and/or electron transport $\left(\eta_{\text{transport}}\right)$. This will have a minimum uncertainty of half of the wavelength step or bandpass (full-width at half maximum, or FWHM) used. Performing IPCE while applying a constant bias often increases the measured photocurrent, which could be due to either a shift of the Fermi level at the CE (not applicable in a three-electrode measurement) or due to increased carrier collection at the WE as a result of an increased depletion width.

7.2 Limitations of the Technique

7.2.1 Confidence in Results

IPCE results are typically quite reproducible between laboratories, but fluctuations in reported data are dependent on the accuracy of the measurements. Large discrepancies between reported IPCE values can be observed when the spectral illumination power density varies from one test setup to another. Therefore, when testing identical samples at different labs, take care to test with similar light intensities at all wavelengths (a nontrivial task). The intensity of the monochromated light can vary widely and is system dependent. Differences in illumination intensities and bandpass at each wavelength should be taken into account when comparing data to the literature values, since nonlinear effects could be seen when the incident light intensity changes significantly. For example, at greater intensities, spontaneous emission from the semiconductor may become a smaller fractional loss (increasing efficiency), while higher currents lead to greater resistive losses (decreasing efficiency). It may be advantageous to monochromate light from a tungsten lamp, since the intense emission lines in the output of most arc lamps can produce spectral artifacts due to nonlinear photocurrent response. One way to help eliminate effects due to low illumination intensity (e.g., unfilled traps preventing efficient carrier extraction) is to apply a white light bias in addition to the monochromated light. The frequency of the modulated monochromatic light must be low enough that the cell has time to respond [1]. For nonlinear cells, the proper bias light level to obtain the 1-sun IPCE is not the 1-sun j_{sc}, but 37 % of the 1-sun j_{sc} (1/2.71828) [2–4]. If bias light is not used, then the measurement assumes that the IPCE is independent of the monochromatic light level.

7.2.2 *Corrosion Potential*

If the samples are chemically unstable in solution in the dark or during illumination, this technique will cause significant damage and complicate the interpretation of any PEC data. Efforts should be made to find an electrolyte that minimizes corrosion during testing. In the case that the electrode is chemically unstable, minimizing the time spent in solution by performing shorter measurements can be beneficial. One way to minimize the photocorrosion of the sample is to start at lower photon energies (long wavelengths) and progress to higher energies (short wavelengths), and swept back to assess any hysteresis which may indicate instability. If photocorrosion is a problem and the signal-to-noise permits, the light intensity may be kept to a minimum by the use of neutral density filters.

7.2.3 *Sources of Error*

Several factors will affect the accuracy of the results. The first and most important concern is with the wavelength-dependent quantification of illumination intensity and the assumptions made therein (see Chapter "Experimental Considerations"). The approximation typically made for photodiodes is that, for a given bandpass of illumination (e.g., 550 ± 10 nm, FWHM = 20 nm), the Gaussian distribution of photon energies is all treated as having one energy. By using the calibration factor for the *peak* of the Gaussian curve (i.e., 550 nm), one will essentially average the calibration factors from 540 to 560 nm to calculate the number of photons. However, if the calibration factors for the wavelengths of interest (i.e., 540–560 nm) differ significantly from that of the peak at 550 nm, some error will be introduced. This experimental error will often also change as a function of wavelength [5]. To minimize this error, one can account for the spectral distribution of the photons at each factory-calibrated wavelength (i.e., use the individual calibration factors to determine the photon rate at every wavelength increment at which the photodiode measures the power), assuming calibration factors are available at wavelength intervals smaller than the bandpass.

Another approach is to use a thermopile with a flat wavelength response over the scanned spectral range and measure the power with a well-calibrated power meter (laser power meters calibrated at specific wavelengths work well) and then convert power into photon flux at all other wavelengths. Errors due to assumption of a constant bandpass intensity at each wavelength are also typically not accounted for and could be significant when a bandpass >10 nm is being used.

Another possible source of error is incorrect filtering of the higher order diffraction peaks generated by the diffraction gratings of the monochromator (e.g., choosing 700 nm produces peaks at $350 - 175$ nm). Appropriate cutoff filters must be used with each grating, and the combination of filter and grating must often be changed several times as one sweeps across the wavelengths of the visible spectrum. An easy way to check for the presence of this error is to simply measure the optical spectrum at each system setting by using a UV–Vis spectrometer. If the error is present, then more than one monochromatic peak will be observed. Broadband stray light increases rapidly with decreasing wavelength and is a significant source of error for wavelengths shorter than 400 nm. Stray light can be minimized by using a double grating monochromator or a bandpass filter in the UV region. Additional sources of error may be related to accurately knowing the true illuminated surface area or the illumination power density. Nonuniformities in the distribution of light across the electrode surface or in the homogeneity of the absorbing sample across the illuminated area (e.g., thickness variations) may also introduce errors in this measurement.

7.3 Pitfalls of the Experiment

Several experimental pitfalls exist, which are dependent on the instrumentation used and its accuracy, including the calibration of the power meter (photodiode) with a NIST traceable source. If photocorrosion is an issue and the samples have a very short lifetime, corrosion currents could interfere with the analysis of the data. Light chopping in conjunction with a lock-in amplifier requires very slow chop rates (<1 Hz) for this experiment, because capacitive charging at the solid–liquid interface may provide large current transients, with time constants up to seconds for some systems, giving erroneously large photocurrent results. These transients, which vary greatly in magnitude and time constant between samples, will throw off the assumption of the lock-in amplifier that the signal is following the square wave of the reference frequency and can produce erroneous results.

7.4 Method

Refer to Section "Cell Setup and Connections for Three- and Two-Electrode Configurations" for a list of minimum equipment required.

Additional components needed for the IPCE experiment are:

- Monochromator and related optics (diffraction grating, slits, focusing assembly, etc.)
- Cutoff filters (for removing higher order peaks from monochromator)
- Second light source (for white light bias, if needed)

In this experiment, the photocurrent (j_{ph}) of the sample is measured versus time (chronoamperometry) with or without an applied bias. In this case, monochromatic light is used over the solar UV–VIS–NIR spectrum (300–1100 nm). Figure 7.1 shows the experimental setup, which consists of an illumination source with a monochromator and a set of filters for order sorting. A mechanical shutter to chop

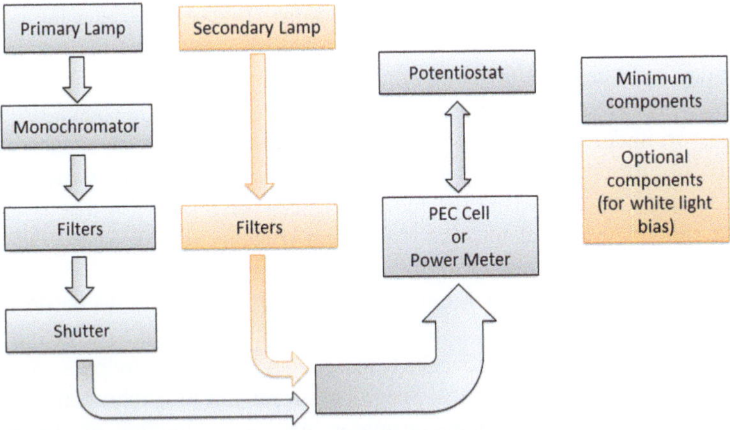

Fig. 7.1 Schematic diagram of a typical experimental setup for measuring IPCE

the beam is preferred. In many cases, this is part of the monochromator and can be controlled independently of the wavelength. If UV light is used, a Quartz cell is required to ensure transmission. Different light sources can be used for this measurement, but the light source must emit photons at the wavelengths at which the samples will be tested (e.g., ~ 300–900 nm). It is also important to have a sufficiently powerful light source (usually 100–1000 W, sample and setup dependent) to achieve an acceptable signal-to-noise ratio during the monochromatic experiments, which themselves have inherently low light intensities due to losses in transmitting the light through the various components of the optical system.

7.4.1 Preparation Time

The preparation time is highly dependent on the type of setup and the sample being used. System calibration prior to sample testing must be performed on a daily basis at the minimum (more often if the optical path lengths are changed) and is often quite time consuming. If the sample area is overfilled with illumination, it is of great importance that the geometric area of the sample exposed to light be known and that the power density at each wavelength be kept constant for all the samples tested.

7.4.2 Calibration of Lamp and Sample Measurement

The intensity of the light beam (P_{mono}) that impinges on the sample should be measured at every pertinent wavelength; this measurement should be performed at least daily after warmup of the lamp ($\sim$30 min). Take care so that the power or power density striking the detector and the sample is constant (or similar) on a day-by-day basis and that there are no changes in the monochromator settings (slit aperture, filter orientation, etc.). Tilting the beam, changing the source or detector distance, or under-illuminating the detector could change the measured light intensity, in turn changing the sample illumination. A suitable calibrated detector should be used for measuring the light intensity (see Pitfalls for discussion); the detector should be placed into the experimental setup *exactly where the sample will reside.*

The settings for the monochromated light experiments need to be determined before the light intensity is measured, and care should be taken to avoid higher order diffraction peaks and to select the proper average bandpass energy for illumination. Consult the technical specification for the monochromator to determine the resolution of the monochromator (e.g., as a function of slit width or number of grating lines). The spectral shape of every type of lamp will be different and depends not only on the optics that focus and filter the light but also on the age of the lamp being used.

The sample should be mounted as in any other PEC performance measurement, which is dependent on the type of sample used, the method for preparing the sample, and the type of measurement cell used. Once setup is complete, a chronoamperometry experiment (current vs. time) should be performed at each calibrated wavelength. It is advisable to initiate the measurement at low photon energy (long wavelengths) and end at high photon energy (short wavelengths).

7.4.3 Applied Bias IPCE Experiment

An applied bias IPCE experiment should be performed in cases where (1) the majority carrier does not possess sufficient energy to drive its respective half-reaction (e.g., HER for n-type and OER for p-type) in a two-electrode measurement or (2) an applied electric field is needed to aid carrier collection. Employing several applied biases provides more insight into sample performance. The experimental setup and procedure remains the same as in the unbiased case, but an additional bias is applied between the WE and RE (three-electrode) or the WE and CE (two-electrode).

7.4.4 White Light Bias IPCE Experiment

A white light bias experiment may be necessary when the intensity of the monochromated light is insufficient to overcome losses due to filling of trap states or spontaneous emission. The additional intensity from a constant background of white light enables carriers generated by the monochromated light to be more efficiently collected in the external circuit.

Equipment modifications are needed for a white light bias (see optional equipment in Fig. 7.1). The experiment requires a low noise broadband DC illumination, while simultaneously illuminating the sample with chopped ($\sim$AC) monochromatic light. The waveform of the chopped monochromic light must be monitored to verify that the chopping speed is low enough for the sample response to reach steady state. The bias level should be set to 37 % of the expected j_{sc} [2–4].

7.5 Measurement and Analysis Time

A complete, wavelength-dependent IPCE measurement and analysis, including lamp warm-up and calibration, can require up to several hours and significant user input to operate both the monochromator and potentiostat. Automation of such systems is possible and can reduce the total required time considerably. The bandpass width, the range of wavelengths to be measured, and the total number of data points collected will also greatly affect experiment time.

7.5.1 Data Analysis

The general equation for calculating IPCE is Eq. 2.5 from Chapter "Efficiency Definitions in the Field of PEC":

$$\text{IPCE}(\lambda) = \text{EQE}(\lambda) = \frac{\text{electrons/cm}^2/\text{s}}{\text{photons/cm}^2/\text{s}} = \frac{\left|j_{ph}(\text{mA/cm}^2)\right| \times 1239.8(\text{V} \times \text{nm})}{P_{\text{mono}}(\text{mW/cm}^2) \times \lambda(\text{nm})}$$

The intensity of the monochromatic light (P_{mono}) is recorded with a power meter equipped with a thermopile detector, calibrated silicon photodiode, or calibrated spectrophotometer (see section Limitations of the Technique); the photocurrent j_{ph} is obtained from the chronoramperometry experiment and usually represents an average steady-state value over a set period of time (at least several seconds). The number of photogenerated electrons can be calculated from the photocurrent using the following.

$$\text{electrons/s} = j_{\text{ph}}(\text{mC/s}) \times 6.241506 \times 10^{15} \text{ electrons/mC} \qquad (7.1)$$

where j_{ph} is the average steady-state photocurrent (mA or mC/s).

Figure 7.2a shows the intensity (as measured by a calibrated silicon photodiode) from a monochromator set to a 20 nm bandpass attached to a 1 kW Xe lamp. The wavelength calibration is set to the center of the bandpass.

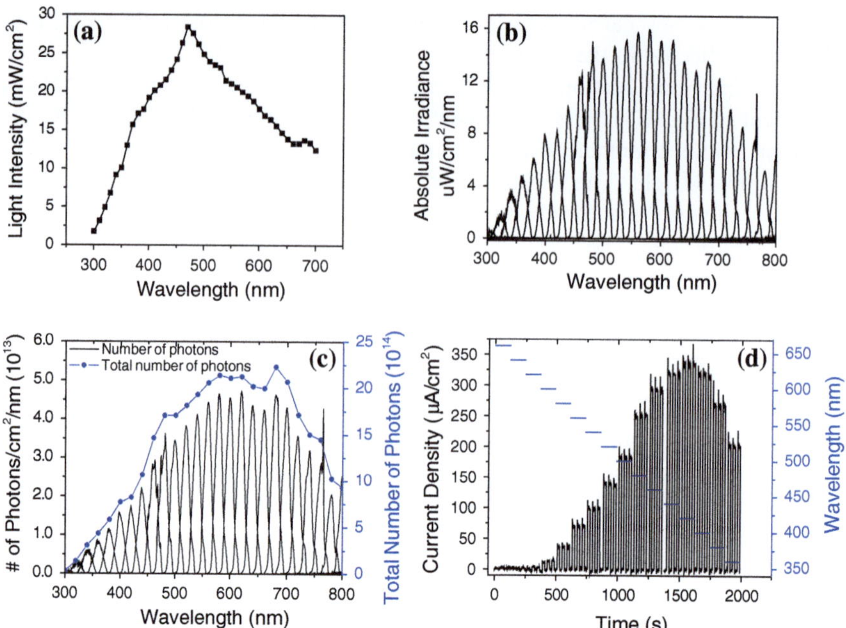

Fig. 7.2 Measured data from an IPCE experiment. (**a**) Total integrated intensity versus wavelength (1 kW Xe lamp) from the output of a monochromator set to a 20 nm bandpass (FWHM) measured by a calibrated silicon photodetector at various wavelengths. For comparison, (**b**) shows the absolute irradiance measurement taken from a similar setup with a calibrated spectrophotometer possessing a resolution of ∼0.5 nm as compared to the silicon detector shown in (**a**). It is worth noting that the spectral distribution at most wavelengths resembles a Gaussian but distortions occur due to specific emission peaks of the light source such as that observed at 764 nm. (**c**) Spectral distribution of photons converted from (**b**) in *black* and the integrated total number of photons in *blue*. (**d**) Current density vs time for an α-Fe$_2$O$_3$ sample. The illumination wavelength is changed periodically and the light is shuttered three times at each wavelength

The energy of a photon is given by Eq. (7.2); the number of photons per J or W × s can be determined at each wavelength by measuring P_{mono} (watts) at λ (nm) as per Eq. (7.3).

$$E(\lambda) = \frac{hc}{\lambda} = \frac{1.988 \times 10^{-13}(\text{mJ} \times \text{nm})}{\lambda(\text{nm})} \text{ per photon} \tag{7.2}$$

$$P_{\text{mono}}(\text{mW}) = \Phi \text{ (photons/s)} \times \frac{1.988 \times 10^{-13}(\text{mJ} \times \text{nm})}{\lambda(\text{nm})} \tag{7.3}$$

$$\Phi \text{ (photons/s)} = 5.03 \times 10^{12}(\text{mJ}^{-1} \times \text{nm}^{-1}) \times \lambda(\text{nm}) \times P_{\text{mono}}(\text{mW}) \tag{7.4}$$

The number of photons at any measured wavelength can be calculated by using Eq. (7.4). For example in Fig. 7.2a, at 470 nm (2.64 eV) the light intensity recorded was 28.55 mW which corresponds to 6.75 × 10^{16} photons. Figure 7.2b shows the spectral distribution of irradiance using a similar lamp and monochromator setup; this measurement was obtained with a calibrated spectrophotometer possessing a resolution of ~ 0.5 nm. The corresponding distribution of photons is obtained using Eq. (7.4) and shown in Fig. 7.2c, and the total number of photons at each wavelength is calculated by integrating the area under the peak. These wavelength-dependent values for the number of photons may vary significantly from lab to lab depending on the full optical setup used and may account, at least in part, for variations in reported IPCE data.

A graph showing the current vs time across various wavelengths is shown in Fig. 7.2d. Note that the frequency of light chopping is sufficiently low to allow the photocurrent to reach steady state to avoid errors from initial capacitance charging. The dark current with the light off should be subtracted from the total current with the light on to produce the photocurrent. The average steady-state photocurrent value is then used in the calculation of IPCE at each wavelength. In this particular measurement, the light is chopped on and off three times at each wavelength to obtain several values for the photocurrent (lock-in amplifiers are generally not recommended due to the relatively long time constants in PEC measurements).

The IPCE can then be determined after determining j_{ph} and P_{mono} at all of the wavelengths. The IPCE of α-Fe_2O_3 is shown in Fig. 7.3a and that of GaP, GaInPN, and $GaInP_2$ are shown in Fig. 7.3b.

For the biased IPCE experiments, data processing is performed in the same manner as for unbiased experiments. Care should be taken to record the applied bias and label the data accordingly such that it is clear that the experiment was performed under bias conditions (see Fig. 7.3a labels). The same rationale should be applied to any measurements performed with a white light bias.

7.6 Photocurrent Spectroscopy

Photocurrent spectroscopy examines the photocurrent produced by an electrochemical cell as a function of wavelength of the incident light. The optical bulk band gap of the semiconductor electrode can be determined along with information about whether it is a direct or indirect transition. Photocurrent spectroscopy provides a direct measurement of the energetic threshold required of photons to produce mobile and extractable charges. This technique is complementary to the optical absorption measured in UV–Vis spectroscopy. Although fundamentally different from UV–Vis due to the inclusion of $\eta_{transport}$ and $\eta_{interface}$ as explained at the beginning of this chapter, photocurrent spectroscopy ideally reveals similar information regarding the band gap and the character of the electronic transition.

The experimental setup and procedure is the same as that required for the basic IPCE experiments except that the requirements of electron counting are not as stringent as with IPCE, as it is not efficiency that is ultimately being reported but

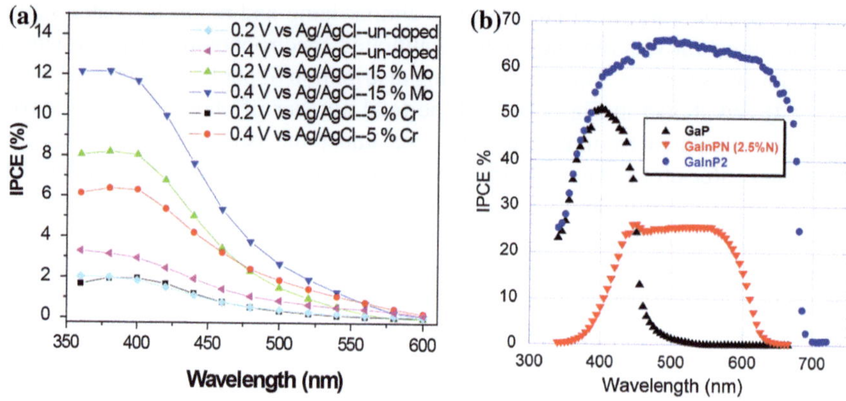

Fig. 7.3 (**a**) Processed IPCE data of doped α-Fe$_2$O$_3$ samples under different applied bias conditions in 1.0 M NaOH. (**b**) Processed IPCE data of GaP, Ga$_{.95}$In$_{.05}$P$_{.975}$N$_{.025}$, and GaInP$_2$ in pH 2 buffer with 1 mM hexaammineruthenium (III) chloride [6, 7]

Fig. 7.4 Allowed direct band gap Tauc plot of a GaInP$_2$ photoelectrode using the normalized photocurrent response. The linear portion of the data is extrapolated to a band gap of 1.81 eV

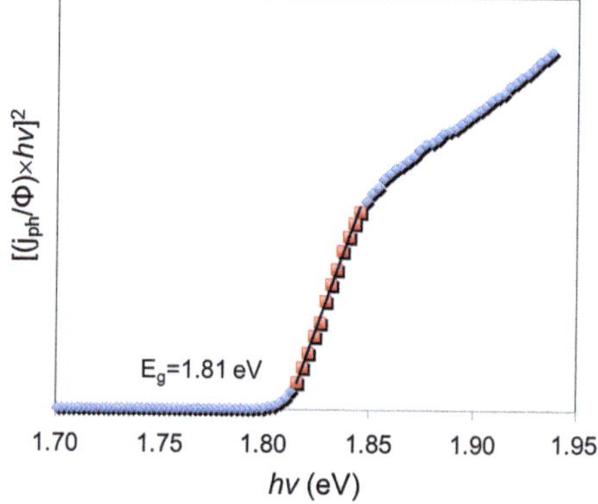

rather the general trend of photoactivity as a function of illumination wavelength. The main difference is in the data analysis. The photocurrent (j_{ph}) can simply be normalized to the photon flux (Φ) at that wavelength from the illumination spectrum and plotted as a function of photon energy. To determine the band gap of the material, a Tauc plot [8] of $(j_{ph}/\Phi \times hv)^n$ is plotted versus hv [9, 10], where n can be 2, 2/3, 1/2, or 1/3. A straight line is fitted at the lower energy region to determine the optical band gap of the material. This procedure is identical to that described in Chapter "UV-Vis Spectroscopy". An example is shown in Fig. 7.4 for a GaInP$_2$ photoelectrode.

References

1. J. Holh-Ebinger, A. Hinsch, R. Sastrawan, W. Warta and U. Wurfel, Dependence of spectral response of dye solar cells on bias illumination, in *Proceedings of the 14th European Photovoltaic Solar Energy Conference and Exhibition*, (2004)
2. T.J. McMahon, K. Sadlon, Errors in calculated air mass 1 short-circuit currents due to non-linear responsivities. Solar Cells **13**, 99–105 (1984)
3. J. Hohl-Ebinger, G. Siefer and W. Warta: Non-linearity of solar cells in spectral response measurements, in *Proceedings of the 22nd European Photovoltaic Solar Energy Conference and Exhibition*, 1CV.2.8 (2007)
4. J. Metzdorf, Calibration of solar cells. 1: The differential spectral responsivity method. *Appl. Optics*, *26*, 1701–1708 (1987)
5. H. Field: Solar cell spectral response measurement errors related to spectral band width and chopped light waveform, in *Proceedings of the 26th IEEE Photovoltaic Specialists Conference*, 1209–1212 (1997)
6. T.G. Deutsch, J.L. Head, J.A. Turner, Photoelectrochemical characterization and durability analysis of GaInPN epilayers. J. Electrochem. Soc. **155**, B903–B907 (2008)
7. Z. Chen, T.F. Jaramillo, T.G. Deutsch, A. Kleiman-Shwarsctein, A.J. Forman, N. Gaillard, R. Garland, K. Takanabe, C. Heske, M. Sunkara, E.W. McFarland, K. Domen, E.L. Miller, J.A. Turner, H.N. Dinh, Accelerating materials development for photoelectrochemical hydrogen production: Standards for methods, definitions, and reporting protocols. J. Mater. Res. **25**, 3–16 (2010)
8. D.L. Wood and J. Tauc, Weak absorption tails in amorphous semiconductors. *Phys. Rev. B*, *5*, 3144–3151 (1972)
9. D.S. Ginley, M.A. Butler, Photoelectrolysis of water using iron titanate anodes. J. Appl. Phys. **48**, 2019–2021 (1977)
10. F.P. Koffyberg, K. Dwight, A. Wold, Interband transitions of semiconducting oxides determined from photoelectrolysis spectra. Solid State Commun. **30**, 433–437 (1979)

Chapter 8
2-Electrode Short Circuit and j–V

Solar-to-hydrogen (STH) conversion efficiency is the most important figure of merit to gage the potential of a semiconductor material to photoelectrochemically split water (see Chapter "Efficiency Definitions in the Field of PEC"). It is projected that STH conversion efficiencies in excess of 10 % will be needed for practical hydrogen production systems [1]. Taken in conjunction with gas detection measurements (see Chapter "Stability Testing"), the photocurrent density (j_{SC}) under short-circuited conditions (i.e., zero applied bias) in a 2-electrode measurement is critical in determining the STH conversion efficiency. Moreover, applied bias experiments using the 2-electrode configuration can shed important light on the water splitting capabilities and limits of a PEC material system.

8.1 Knowledge Gained

A 2-electrode current density–voltage (j–V) experiment, where bias is varied between the WE and CE, can be used to obtain an applied bias photon-to-current efficiency (ABPE) under AM 1.5 G illumination. In the special short-circuited case (i.e., zero applied bias), STH can be derived for viable water splitting materials/devices. For materials that cannot split water by themselves but require only a moderate (0.5 V) bias to achieve photoelectrolysis, integration of the material into a multibandgap tandem structure may yield a viable device. On the other hand, if the material system can split water, but only needs a small bias (0.1 V) to maximize efficiency, one might instead choose to modify the semiconductor band gap, apply a surface catalyst, alter the cell geometry, or enhance the electrolyte conductivity. If the required bias is too high (1.5 V), the photoelectrode may require major changes to its structure, composition, morphology, or synthesis conditions.

Z. Chen et al., *Photoelectrochemical Water Splitting*,
SpringerBriefs in Energy, DOI: 10.1007/978-1-4614-8298-7_8,

8.2 Limitations of Technique

8.2.1 Credibility of Results

A 2-electrode experiment can yield a true STH efficiency if a carefully calibrated light source is used and if gas evolution is verified. If every photogenerated electron and hole drive the HER and OER, respectively (known as 100 % Faradaic efficiency) the measured photocurrent density can directly correlate to STH.

8.2.2 Corrosion Potential

Short-circuit measurements mimic real-world conditions where photogenerated electrons and holes may have myriad reduction and oxidation half-reactions (besides HER and OER) that can damage the electrode surface. In addition, the photoelectrodes may simply be prone to chemical instability in aqueous environments with acid or base. The damage can be minimized by limiting the duration of the measurements.

8.2.3 Sources of Error

The incorrect determination of illuminated surface area will directly affect measured j_{SC}. Photoelectrons and holes that participate in side reactions such as corrosion, oxygen reduction, or the oxidation of organic contaminants/surfactants can lead to inflated estimates of STH from this technique. Failure to prevent the product of one electrode (H_2 or O_2) from reaching the complementary electrode (and consequently undergoing the reverse reaction) can inflate the STH. Inadequate calibration of the light source can lead to over or underestimated STH. An electrolyte with low ionic conductivity will reduce j_{SC} by increasing series resistance.

8.3 Method

8.3.1 Experimental Setup and Procedure

Either a low impedance ammeter or a potentiostat can be used for this measurement. Connect the WE lead to the semiconductor and the CE lead (which is shorted to the RE lead if available) to an appropriate counter electrode. Immerse the two electrodes in an appropriately sparged electrolyte (see Section "Cell Setup and Connections for 3- and 2- Electrode Configurations" . Ideally, the two electrodes are inserted into separate compartments which are independently sparged with the respective gaseous product (H_2 for the cathode and O_2 for the anode). The compartments are connected by a separator (e.g., porous frit or an appropriate proton/hydroxyl exchange membrane. Take care that the H_2 from the cathode does not migrate to the anode, and that the O_2 from the anode does not migrate to the cathode, or else the j_{SC} will become inflated. In the absence of a 2-compartment cell, ensuring that these products do not cross over takes top priority, and the solution should be actively sparged with an inert gas such as Ar or N_2.

Illuminate the WE at AM 1.5 G and measure the current at zero-bias as a function of time. Depending on the material, a few seconds to a few minutes may be needed for the current to stabilize. Dark current can be determined by intermittently blocking the light source with a shutter or chopper.

8.3.2 2-Electrode j–V Measurement

For the 2-electrode current density–voltage (j–V) measurement, an external applied bias (using a source meter or a potentiostat) is applied between the WE and CE. The optimal direction and endpoints of the voltage scan are dependent on the material conductivity. A scan rate in the range of 10–100 mV/s is typically used for measurement. See Section "Cell Setup and Connections for 3- and 2- Electrode Configurations" for additional details on setup. The 2-electrode j–V experiment can be performed in the dark and then under illumination, or it can be carried out with chopped light. Figure 8.1 gives an example of a chopped-light 2-electrode j–V plot.

8.3.3 Preparation and Measurement Time

The lamp usually needs 30 min to stabilize, and sparging requires at least 10 min, depending on the cell configuration. Short-circuit current, including short-term trends, can be measured in a few minutes, but long-term testing is essential to monitor the stability of the material under study (see Chapter "Hydrogen and Oxygen Detection from Photoelectrodes"). Long-term testing can continue for a matter of hours, or even days, in which case continued sparging may be difficult to maintain due to evaporation of the electrolyte.

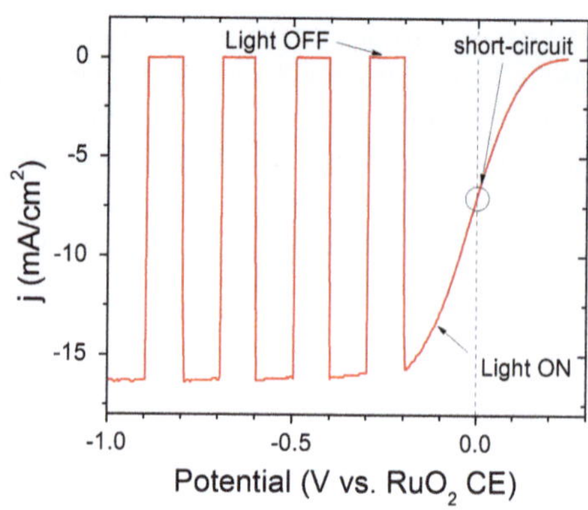

Fig. 8.1 Chopped-light 2-electrode j-V curve for a GaInP$_2$/GaAs tandem cell with a RuO$_2$ counter electrode in 3 M H$_2$SO$_4$ under AM 1.5 G illumination

8.3.4 Data Analysis and Expected Results

8.3.4.1 2-Electrode Short-Current Density Measurement

Data analysis is straightforward for this method. The steady-state j_{SC} from a photoelectrode can be used to calculate STH using Eq. 2.2 from Chapter "Efficiency Definitions in the Field of PEC":

$$\text{STH} = \left[\frac{\left| j_{SC}(\text{mA/cm}^2) \right| \times (1.23\,\text{V}) \times \eta_F}{P_{\text{total}}\,(\text{mW/cm}^2)} \right]_{\text{AM1.5\,G}}$$

The light source should be tuned to have an output which matches the AM 1.5 G spectrum at one sun intensity (100 mW/cm^2) to most accurately estimate real-world STH efficiencies. An example is shown in Fig. 8.2 for a GaInP$_2$/GaAs tandem cell. Note that this particular sample is not the same sample shown in Fig. 8.1.

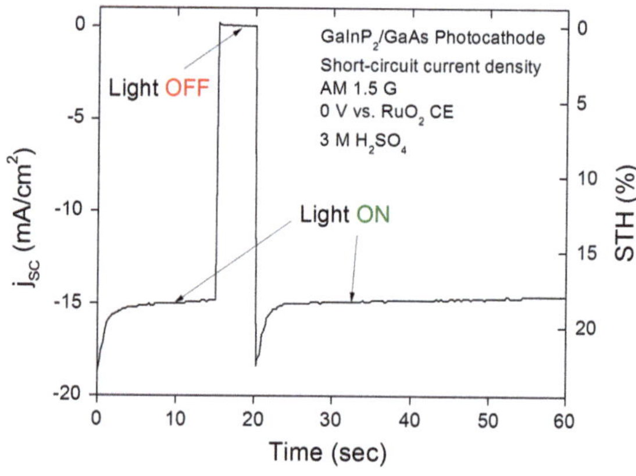

Fig. 8.2 Short-circuit current density of a GaInP$_2$/GaAs photocathode connected to a RuO$_2$ CE in 3 M H$_2$SO$_4$ under 1 sun AM 1.5 G illumination from a tungsten lamp calibrated to a 1.8 eV GaInP$_2$ reference cell. The corresponding STH reaches over 17 %, assuming $\eta_F \sim 1$

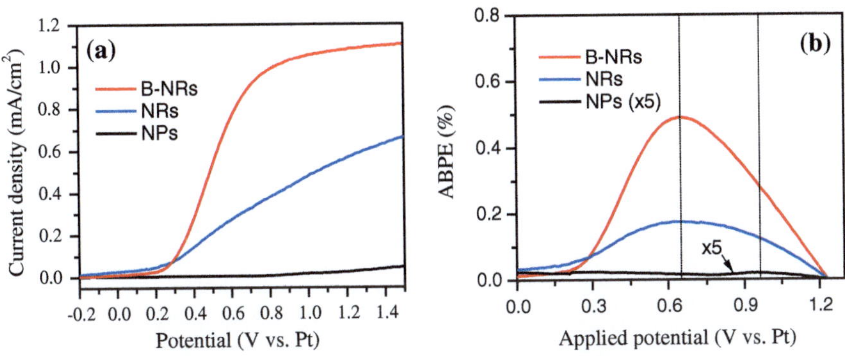

Fig. 8.3 (**a**) 2-electrode j–V of several nanostructured TiO$_2$ photoanodes with a Pt CE, including nanoparticles (NPs), nanorods (NRs) and branched nanorods (B-NRs) [2]. (**b**) The corresponding ABPE plot generated using Eq. 2.3 in Chapter "Efficiency Definitions in the Field of PEC". The maximum ABPE of 0.49 % for the B-NRs occurs at a V_b of 0.65 V versus Pt, whereas the maximum ABPE of 0.02 % for the NPs occurs at a V_b of 0.94 V versus Pt

8.3.4.2   2-Electrode j–V Measurement

The applied bias photon-to-current conversion efficiency (ABPE) can be determined using Eq. 2.3 from Chapter "Efficiency Definitions in the Field of PEC":

$$\text{ABPE} = \left[\frac{\left| j_{ph}\,(\text{mA/cm}^2) \right| \times (1.23 - |V_b|)\,(\text{V}) \times \eta_F}{P_{total}\,(\text{mW/cm}^2)} \right]_{\text{AM1.5 G}}$$

An example of a 2-electrode j–V measurement and the corresponding ABPE plot is shown in Fig. 8.3 for several nanostructured TiO$_2$ photoanodes.

References

1. Multi-Year Research, Development, and Demonstration Plan, *Hydrogen Production*, (2012), http://www1.eere.energy.gov/hydrogenandfuelcells/mypp/pdfs/production.pdf
2. I.S. Cho, Z. Chen, A.J. Forman, D.R. Kim, P.M. Rao, T.F. Jaramillo, X. Zheng, Branched TiO$_2$ nanorods for photoelectrochemical hydrogen production. Nano Lett. **11**, 4978–4984 (2011)

Chapter 9
Hydrogen and Oxygen Detection
from Photoelectrodes

The chemical products of PEC water splitting processes are the evolved hydrogen and oxygen gases. Standard experimental methods for detecting and validating the quantity and quality of the product gases are critical.

9.1 Knowledge Gained

In PEC experiments, measured photocurrents reflect consequences of the electrochemical reactions, including desired and undesired product formation and the decomposition and dissolution of electrode materials. The photocurrents, thus, cannot be used directly for quantitative analyses of the desired reaction (although they do provide a good first-order approximation once gas evolution has been verified). In PEC water splitting, it is quantitative detection of H_2 and O_2 in the gas phase that provides direct evidence for true water splitting, a primary requirement for the validity of solar-to-hydrogen (STH) efficiency measurements (see Chapter "Efficiency Definitions in the Field of PEC"). Calibrated gas analyzers, such as a gas chromatograph (GC) or mass spectrometer (MS), can be used to determine the nature and extent of the photoelectrochemical reactions. The products can be monitored over the course of the reaction, confirming turnovers of the electrocatalysis (i.e., moles of the product gases per active sites), while at the same time current can be monitored in the external circuit. From these two values, Faradaic efficiencies for the water splitting reaction in the given system can be calculated.

9.2 Method

Conventional analysis for evolved gases from an electrochemical system can be performed by using the generalized reactor cell as shown in Fig. 9.1 [1, 2]. In this method, the evolved gases are quantified simply by measuring the volume of the

Z. Chen et al., *Photoelectrochemical Water Splitting*,
SpringerBriefs in Energy, DOI: 10.1007/978-1-4614-8298-7_9,
© The Author(s) 2013

Fig. 9.1 Schematic of a
conventional 2-compartment
PEC cell with a headspace
that allows for independent
measure of the gases evolved
at each electrode

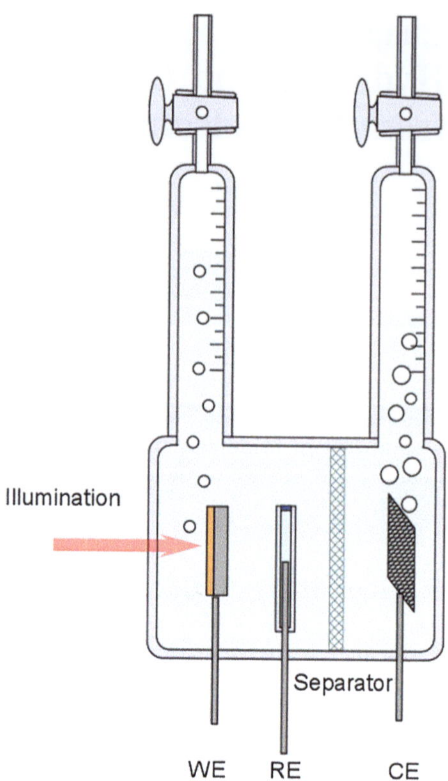

gases separately for the anode and cathode. The identities of the stored gases must
still be verified ex-situ by GC or MS.

GC analysis is usually used for the detection of H_2 and O_2, as well as other
gases that originate from undesired reactions or electrode decomposition. GC
equipment consists of a gas injection component and a gas separation component
with suitable column(s), followed by detector(s). A schematic image is shown in
Fig. 9.2.

For gas injection, a conventional method is to fill a syringe with a known
amount of gas product. Alternatively, a multi-valve setup with a sampling loop can
be used for injection. Micro-GCs equipped with TCDs are often useful because of
their high sensitivity, requiring very small volumes of gases for analysis. Micro-
GC equipment often utilizes pumping-assisted injection from the gas stream.
Columns of molecular sieves 5A, 13X, or HayeSep A are effective for H_2, O_2, and
N_2 separation. The particle size of the adsorbate as well as the diameter and length
of the columns should be taken into consideration for good separation. These
columns strongly adsorb H_2O and CO_2, and the regeneration of the column by
baking (removal of the adsorbed species) is recommended periodically when the
retention times become too short for adequate separation.

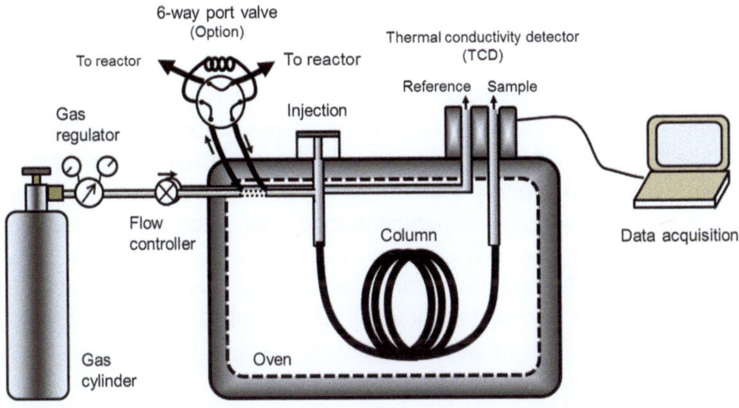

Fig. 9.2 Schematic representation of a basic GC equipped with a thermal conductivity detector

A GC equipped with one or two thermal conductivity detectors (TCDs) is commonly used for both qualitative and quantitative gas detection in catalytic processes. TCDs measure the differences in thermal conductivity of the injected gases relative to the carrier gas, and therefore are suitable for inorganic gases like H_2, O_2, and N_2. TCDs are universal, stable, and moderately sensitive, and are usually used with He carrier gas. However, using He carrier gas precludes the analysis of H_2 in a gas mixture. The thermal conductivity of H_2 is too close to that of He, resulting in irregular peak shapes (often with a W-shape) and thus quantitative results are not possible. Ar is a suitable carrier gas for H_2 detection. When high sensitivity for O_2 detection is also needed, a second TCD (if applicable) with He carrier gas may be employed for detecting N_2 and O_2.

Although MS possesses high sensitivity and is useful for qualitative analysis of small amounts of gases [3], it lacks the quantitative accuracy of GC because of the difficulty in the constant introduction of gases into the detector parts, which must be kept under high vacuum. Quantitative analyses of small amounts of target gases by MS require careful calibration by controlling the pressure of each target component. DEMS (differentially electrochemical MS) is often used in the field of electrochemical measurements for detection of small amounts of gas products from the surface monolayer of the electrode [4–7]. It is a strong tool, especially for the detection of isotopic products for mechanistic studies including kinetic isotope effects, isotope exchange, and isotope tracing experiments. Currently, there are limited examples of applying DEMS to photoelectrochemistry [8], but the approach is certainly feasible. For isotope experiments, one can also utilize a GC-TCD-MS (combination of TCD and MS detections for quantitative and qualitative analyses, respectively) [9]. In this section, the experimental procedure for gas measurement with in-line GC-TCD is described.

9.3 Experimental Procedure

Conventional PEC electrodes can be used for the measurement. Current–voltage and current–time curves are measured using a conventional electrochemical cell equipped with a planar window and a standard potentiostat, as described in Section "Cell Setup and Connections for 3- and 2- Electrode Configurations". Depending on the particular experiment, one may choose to operate a 3- or 2-electrode setup. For true STH efficiency measurements, a zero-bias, 2-electrode setup must be used. The electrodes are placed in the electrochemical cell which can be isolated from the environment by sealed joints (with grease or O-ring connections, see Fig. 3.1 in Section "Cell Setup and Connections for 3- and 2- Electrode Configurations"). The leaking of ambient air into the system and product gases out of the system interferes with accurate quantification.

The particular configuration of the PEC cell is determined by the type of light source (discussed below) as well as the rates of gas production from water splitting (lower rates may require smaller volume cells). Figure 9.3 shows some examples of PEC reactors, including a (a) batch, (b) flow and (c) recirculating batch reactor.

9.3.1 Batch Reactor

Batch reactors (Fig. 9.3a) are simple closed systems, which may be suitable for systems with a low rate of gas production. Any size of the reactor can be used, but smaller reactors enable higher concentration of gaseous products due to their

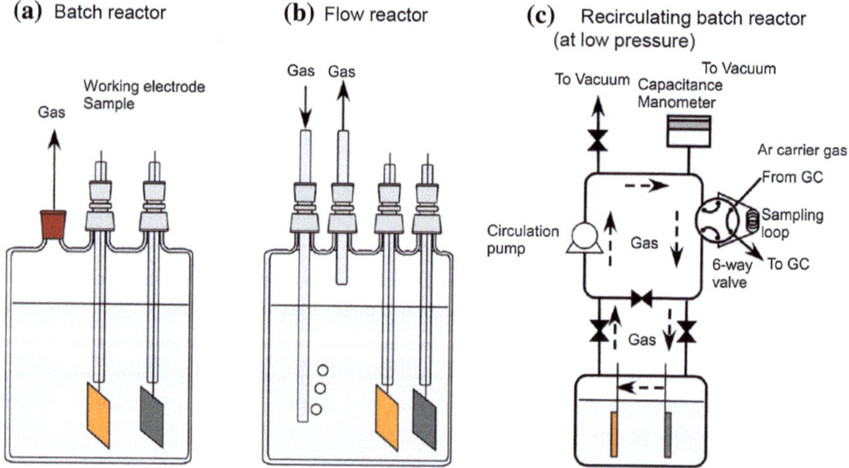

Fig. 9.3 Schematic images of examples of photoelectrochemical reactors; **(a)** Batch reactor, **(b)** Flow reactor, **c** Recirculating batch reactor (evacuation is optional)

smaller volume, which allows for higher signal. Prior to performing measurements, the system should be purged thoroughly with inert gas such as N_2 or Ar to remove the remaining and dissolved gases in H_2O. A GC-TCD, either in-line or ex-line, can be used for gas analyses by taking a known volume of the gas from the reactor, using a gas-tight syringe or a sampling valve. The evolved gases are accumulated in the reactor with time, so there may be increased likelihood back-reaction (conversion of H_2 and O_2 to form H_2O) in the case of a single vessel cell.

9.3.2 Flow Reactor

The second example shown in Fig. 9.3b is a flow reactor, which is useful especially when a highly sensitive analyzer (e.g., micro-GC) is available. A flow rate of the sweep gas should be tuned based on the (expected) evolution rate of H_2 and O_2 as well as the sensitivity of the detector used. Typically, reactors in the range of 50–200 cm^3 are best matched with a flow rate of sweep gas of 5–20 cm^3 min^{-1}. The flow rate of the sweep gas should be compared to the dead volume of the system for accurate measurements. Argon may be used as a carrier gas and is beneficial for detecting not only H_2, but also N_2 to assess any leakage of air into the system or to monitor corrosion of nitrogen-containing photoelectrode materials. Otherwise, N_2 can be applied as a carrier gas. Before the measurements, thorough purging of the reactor is required to remove all the dissolved gases in H_2O.

It is useful if the GC is directly connected to the exit of the reactor (in-line GC), so that a direct injection of the sample can be achieved by a sampling valve. In the case of a micro-GC, the gas is typically sampled from the line using a micro-pump, and thus one should be aware of the pumping rate (typically ~ 5 cm^3 min^{-1}) compared to the flow rate of the sweep gas, to make sure that the sample is appropriately injected. Calibration of the GC-TCD should be conducted using a known concentration of H_2 and O_2. By introducing controlled flow rates for each gas (H_2, O_2, and a sweep gas), variation of the partial pressures can be achieved. In principle, changes in gas volume (partial pressure) due to the reaction (water splitting increases the number of molecules in gas-phase) can be accurately corrected using an internal standard, e.g., using a known flow rate of N_2 (internal standard) which can also be diluted with Ar (when Ar is used as a carrier gas for GC). However, this volume change is negligible if the products are at low concentration levels.

9.3.3 Recirculating Reactor (in Vacuum)

The third example of the reactor is a closed recirculation system, as displayed in Fig. 9.3c. The gases in the system (typically, 300–1,000 cm^3) can be recirculated by a circulating pump with a speed >3 cm^3/s to attain homogeneous gas mixing in the system. All valves are made of grease-less valves, or of Pyrex glass sealed with

Apiezon vacuum grease, and appropriately placed to direct recirculating gases. Glass valves sealed with PTFE are not suitable for this measurement because they allow small amounts of air into the system during operation. The system can be evacuated by a mechanical rotary pump, in which case, a liquid-N_2 cooled trap is placed before the rotary pump to avoid back diffusion of oil and to remove water and other impurities that would otherwise damage the pump. Before performing measurements, the headspace of the cell is directly evacuated with the vacuum line to clear the atmosphere of air (O_2) and to remove the dissolved gases from the electrolyte. The headspace is then refilled with the same inert gas as the carrier gas in the GC. The evacuation cycling should be repeated several times for complete removal of all dissolved gases. During this process, some of the water may be evacuated, but the loss should be negligible compared with what remains in the cell. The PEC measurement can be carried out under vacuum (vapor pressure of $H_2O \sim 2$ kPa) or a small amount of inert gas (e.g., Ar ~ 4–10 kPa) as a carrier gas. A vacuum environment is preferred for ease of releasing the evolved gases. A leak test of the system should be carefully performed by monitoring the pressure rise when kept under vacuum. Alternatively, if carrier gases such as Ar or He are used, detection of N_2 while sampling will indicate that air has leaked into the cell.

Calibration of the GC-TCD should be conducted by introducing known quantities of H_2, O_2, and N_2 prior to measurement. For the calibration, several measurements should be made at pressures in the actual range expected to be reached during the PEC measurement. By knowing the total volume of the system and the partial pressure of the products determined by a GC-TCD, the amount of evolved gases can be rigorously quantified. To measure the total volume of the system, a container with known volume can be attached to the system, and the pressure can be monitored to determine the system volume by using Boyle's law.

After driving the PEC reaction, the gas is sampled at a certain timepoint, typically at least 10–15 min. For the use of a conventional GC-TCD, the sampling loop (typically 0.5–2 cm^3) is filled with the gas in the system by controlling the 6-way valve. Then, gas in the sampling loop is injected to the GC by switching the 6-way valve. The gas in the system is continuously circulated by a circulation pump (e.g., Makuhari glass, MR-2000) during the measurement to remove gas concentration gradients.

9.4 Limitations of Technique

This technique is limited to detection of the evolved gases from a PEC experiment. After illumination, the electrode should be analyzed for signs of corrosion and the electrolyte should be separately analyzed to evaluate the presence of any corrosion products. For the separate detection of gases from the anode and cathode, the cell should ideally be divided into two compartments using a separator such as a fine pore glass frit or ion exchange membrane (Fig. 9.2b) to prevent product crossover which can be a major limitation in single compartment cells.

9.4.1 Sources of Error

Leakage of ambient air into the system is a serious problem because O_2 present in air will produce erroneously high readings which could be erroneously attributed to O_2 evolved from the anode, in the case of water splitting. Leakage should be continually evaluated by employing leak detection tools (helium leak detector or discharge by a Tesla generator), or by observing increases in pressure when no reaction is occurring. In addition, leakages can result in loss of product and produce an underestimate of electrode performance.

Calibration of the GC should be carefully performed within the appropriate range of concentrations and pressures for each gas component. In the case of the flow reactor, the gas flow rates should be carefully calibrated before calibrating the GC itself. In the case of the recirculating reactor in vacuum, the volume of the system should be calibrated by using a vessel or a container with a known exact volume. To be strict, the temperature of the system should also be kept constant during and between each measurement. It is thus useful to utilize water baths or cooling systems to maintain constant temperatures during the PEC measurements, which are typically carried out at room temperature.

9.5 Examples

The first example demonstrates the measurement for water splitting to H_2 and O_2 using a $IrO_2/TaON$ electrode under visible-light irradiation with a 300 W Xe lamp [10]. A flow reactor (Fig. 9.3b) was employed with a single vessel cell (Fig. 9.3a), with Pt wire as a CE in 0.1 M Na_2SO_4 aq. (pH 6). The evolved gases were characterized by a GC-TCD (Agilent MicroGC G3000, MS-5A, TCD, Ar carrier), which was directly connected to the photocatalytic reactor system. As can be seen in Fig. 9.4, the amount of evolved H_2 and O_2 increased monotonically with time at the nearly stoichiometric ratio of 2:1 expected from water splitting. The low amount of H_2 evolved compared with expected amount based on half of the electrons passing through the outer circuit ($e^-/2$, shown as a broken line) indicates that the Faradaic efficiency (η_F) is not 100 %, and that there may be some product crossover between the electrodes. Also, the 10 min delay before the onset of H_2 and O_2 was likely due to self-oxidation of the TaON surface in the initial period [10].

In the same manner, the PEC performance for water splitting has also been reported for an IrO_2 deposited GaN:ZnO electrode [11]. The system used was a recirculating reactor (Fig. 9.3c) with a single vessel cell (Fig. 9.3a). Over the full irradiation period (23.5 h), 5.9 µmol of O_2 and 12.2 µmol of H_2 were evolved on the working and counter electrodes. The number of electrons (ca. 24 µmol) calculated from the amount of evolved H_2 was smaller than the electrons calculated by the measured photocurrent (33 µmol). This is most likely because of back reaction, which should be prevalent on the platinum counter electrode before gas detection by the GC.

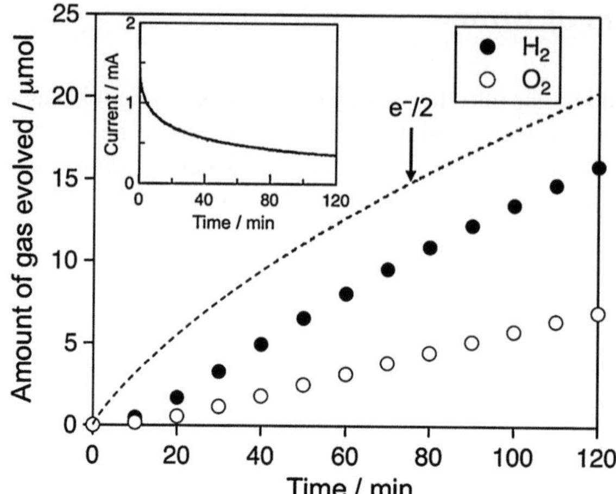

Fig. 9.4 Time evolution of PEC water splitting on a IrO_2/TaON electrode under visible light (of wavelength longer than 420 nm) illumination using a 300 W Xe lamp (Cermax; equipped with a cold mirror: CM2, a cut-off filter: Hoya L42) at 0.8 V versus Pt in an aqueous 0.1 M Na_2SO_4 electrolyte at 288 K. Stoichiometric H_2:O_2 (2:1) gas production was observed. Adapted from Ref. [10]

9.6 Data Analysis

The peak area given by the TCD signal correlates linearly with the concentration of each gas. Based on the calibration for the GC-TCD, the area of the H_2 peak can be converted to mol H_2. When divided by the measurement time, the rate of H_2 evolution is obtained. (Area under H_2 peak) × (Conversion factor mol H_2/area)/ Time (s) = (rate mmol H_2 s^{-1}). This rate is then used with Eq. 2.1 of "Efficiency Definitions in the Field of PEC" to determine the STH.

References

1. A. Fujishima, K. Honda, Electrochemical photolysis of water at a semiconductor electrode. Nature **238**, 37–38 (1972)
2. K. Fujii, T. Karasawa, K. Ohkawa, Hydrogen gas generation by splitting aqueous water using n-type GaN photoelectrode with anodic oxidation. Jpn. J. Appl. Phys. **44**, 543–545 (2005)
3. O. Khaselev, J.A. Turner, A monolithic photovoltaic-photoelectrochemical device for hydrogen production via water splitting. Science **280**, 425–427 (1998)
4. H. Baltruschat, Differential electrochemical mass spectrometry. J. Am. Soc. Mass Spectrom. **15**, 1693–1706 (2004)
5. O. Wolter, J. Willsau, J. Heitbaum, Reaction pathways of the anodic oxidation of formic acid on Pt evidenced by ^{18}O labeling—a DEMS study. J. Electrochem. Soc. **132**, 1635–1638 (1985)

6. S. Wasmus, S.R. Samms, R.F. Savinell, Multipurpose electrochemical mass spectrometry: a new powerful extension of differential electrochemical mass spectrometry. J. Electrochem. Soc. **142**, 1183–1189 (1995)
7. M. Fujihara, T. Noguchi, A novel differential electrochemical mass spectrometer (DEMS) with a stationary gas-permeable electrode in a rotational flow produced by a rotating rod. J. Electroanal. Chem. **347**, 457–463 (1993)
8. P. Bogdanoff, N. Alonso-Vante, on-line determination via differential electrochemical mass spectroscopy (DEMS) of chemical products formed in photoelectrocatalytical systems. Berichte der Bunsengellschaft fur physikalische Chem. **97**, 940–943 (1993)
9. K. Takanabe, E. Iglesia, Mechanistic aspects and reaction pathways for oxidative coupling of methane on $Mn/Na_2WO_4/SiO_2$ catalysts. J. Phys. Chem. C **113**, 10131–10145 (2009)
10. R. Abe, M. Higashi, K. Domen, facile fabrication of an efficient oxynitride TaON photoanode for overall water splitting into H_2 and O_2 under visible light irradiation. J. Am. Chem. Soc. **132**, 11828–11829 (2010)
11. H. Hashiguchi, K. Maeda, R. Abe, A. Ishikawa, J. Kubota, K. Domen, Photoresponse of GaN:ZnO electrode on FTO under visible light irradiation. Bull. Chem. Soc. Jpn. **82**, 401–407 (2009)

Chapter 10
Stability Testing

10.1 Knowledge Gained

Photocorrosion in aqueous environments is one of the most significant obstacles to the widespread deployment of semiconductor materials as PEC devices for solar hydrogen production. The photogenerated holes and electrons in semiconductor electrodes are generally characterized by strong oxidizing and reducing potentials, respectively. Instead of driving the OER or HER, these holes and electrons may oxidize or reduce the semiconductor itself, causing undesired physical and chemical changes.

The photocurrent durability test outlined here involves evaluating the current density versus time under illuminated operational conditions. For a more thorough understanding of the nature of any corrosion processes occurring, additional physical and chemical characterization, such as optical or electron microscopy, X-ray photoelectron spectroscopy (XPS), X-ray diffraction (XRD), Raman spectroscopy, etc., should be carried out before and after stability testing. The long-term stability tests of PEC materials under illumination are often performed with an applied bias between the WE and CE to simulate water splitting conditions (since most materials studied to date cannot split water spontaneously). A bias is chosen that results in photocurrent densities consistent with anticipated PEC operations (typically 1–10 mA/cm^2). It is recommended that the stability test be performed at multiple biases, if possible, in order to gain deeper insight into the stability behavior of the material system under various operating conditions.

10.2 Limits of Experiment

Stability testing is a relatively straightforward procedure. However, there are a few experimental details to note that can help to minimize sources of error.

First, lamp stability and heating of the cell may become a problem for stability experiments carried out over an extended period of time. The light source should

Z. Chen et al., *Photoelectrochemical Water Splitting*,
SpringerBriefs in Energy, DOI: 10.1007/978-1-4614-8298-7_10,

be set to simulate the AM 1.5 G spectrum and intensity using a calibrated reference cell, if possible. However, the output may need to be checked periodically (at least once every 24 h), and the bulb may even need replacement for tests that last for hundreds of hours. Active cooling (e.g., by using a water-jacketed cell) may help manage temperature fluctuations.

Second, although it is best to actively sparge an electrolyte to prevent product crossover and to ensure that an electrode is in an environment saturated with the product it evolves (H_2 or O_2), the electrolyte is generally not sparged during a stability test because of the need to humidify the sparged gas for long-term testing to prevent changes in electrolyte concentration over time from solvent evaporation. If a surfactant is added to the electrolyte to minimize the buildup of bubbles on the electrode surfaces, sparging is also problematic because it causes foaming in the solution. However, without sparging, product crossover can occur, giving rise to a current that is not due to water splitting and may result in a false, higher current reading. A 2-compartment cell may help to mitigate crossover.

10.3 Method

The standard procedure for stability testing utilizes the 2-electrode short-circuit measurement detailed in Chapter "2-Electrode Short Circuit and j-v ". See Section "Cell Setup and Connections for 3- and 2- Electrode Configurations" for a discussion on basic cell setup and electrolyte selection.

A good starting point is to evaluate the current density versus time at zero-bias for 1 h. However, since many materials are unable to split water spontaneously, an external applied bias may be applied. If the semiconductor appears to be reasonably stable for 1 h, this experiment needs to be repeated for longer testing times, i.e., 24 h or longer. The aqueous solution can be analyzed for corrosion products which may have formed during the experiment and the electrode surface can be characterized physically and chemically for signs of corrosion. For example, for $GaInP_2$, one may visibly observe pitting on the surface of the electrode. Inductively coupled plasma mass spectrometry (ICP-MS) can measure the gallium concentration in solution, while a profilometer or optical microscope can be used to characterize the surface morphology. Repeat the experiment with a different electrode in an electrolyte solution with a different pH, as applicable for the material type.

10.4 Data Analysis and Expected Result

The result of this test is a plot of current density (j_{SC}) versus time, indicating the stability of the photoelectrode under simulated water splitting conditions. A constant photocurrent density (cathodic current for hydrogen production and

Fig. 10.1 Short-circuit zero-bias current density versus time plots for two GaPN (3 %): p-Si tandem cells (identified A and B) in sparged 3 MH_2SO_4 under AM 1.5 G illumination (set using a 250 W tungsten halogen lamp and a calibrated GaPN reference cell) for 1 h. Sample B exhibits a high initial photocurrent that quickly decreases, whereas sample A exhibits lower photocurrent initially but is more stable over time [1]

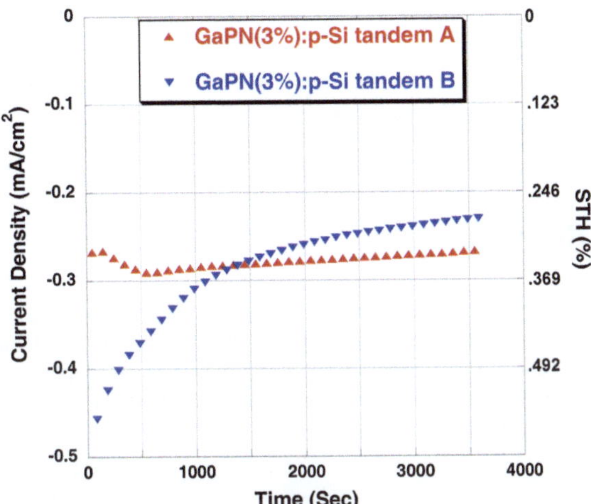

anodic current for oxygen production) over a duration of 1 h would indicate that the electrode shows initial stability and may warrant a prolonged stability test. A decrease in photocurrent density could result from a number of complex bulk and/or surface degradation pathways and will not be discussed in detail here. However, performance loss is often the result of some form of corrosion of the electrode surface in contact with the electrolyte.

An example of a stability plot obtained from a 2-electrode test under zero-bias and AM 1.5 G illumination is shown in Fig. 10.1. The photocurrent density generated from the GaPN (3 %): p-Si tandem cells (GaP$_{.97}$N$_{.03}$ grown on a p/n Si substrate) is in the range of 0.2–0.3 mA/cm^2 (0.3 % STH efficiency), although both cells exhibit degradation in the current density (toward zero) over time.

The current density in Fig. 10.1 is negative (cathodic), indicating that hydrogen production is occurring at the WE surface of the tandem cell. Tandem cell A appears to produce less current initially, but is relatively more stable than Tandem cell B, which possesses higher initial activity, but exhibits a more rapid decay such that it becomes less active than Tandem Cell A after 1,300 s. It should be noted that there is no general curve shape expected for the stability test and that it is highly material/system dependent.

Extrapolating the current density versus time to zero current on the time axis can give a rough estimate of expected lifetime of the electrode. For example, if one makes a linear fit to the rate of decline for tandem cell B in the region >1500 s, then the resulting slope of 4×10^{-5} mA/cm^2/s estimates a total lifetime of 2 h. Other fits to the data may be more appropriate, and presumably should reflect the physical/chemical mechanism of corrosion and is subject to interpretation.

If the electrode is stable for 1 h, durability testing should be repeated for longer testing times, e.g., 24 h, 200 h, 1000 h, and eventually toward 5000 h (see Chapter "PEC Characterization Flowchart") for a more conclusive determination of stability. A simple extrapolation to determine electrode lifetime (as done above) is not always accurate; it is not uncommon to observe sharp, and unexpected, drops in stability. As the field of PEC materials design advances and more stable materials are created, more sophisticated accelerated stability tests will need to be designed and implemented.

Reference

1. Z. Chen, T.F. Jaramillo, T.G. Deutsch, A. Kleiman-Shwarsctein, A.J. Forman, N. Gaillard, R. Garland, K. Takanabe, C. Heske, M. Sunkara, E.W. McFarland, K. Domen, E.L. Miller, J.A. Turner, H.N. Dinh, Accelerating materials development for photoelectrochemical hydrogen production: standards for methods, definitions, and reporting protocols. J. Mater. Res. **25**, 3–16 (2010)

Glossary

α	Absorption coefficient. The absorbance normalized to the pathlength through the sample (typically thickness).
ε	Molar absorptive coefficient of the material $(M^{-1}cm^{-1})$. A term used in Beer's Law that is related to the amount of light absorbed for a given sample.
ε_s	Static relative permittivity of the semiconductor material. A term that measures the amount of stored electrical energy for a given material when a potential is applied. Also referred to as the static dielectric constant $\varepsilon_s = \varepsilon_r \varepsilon_o$.
ε_r	Relative permittivity for a semiconductor. Typical values for single crystals is 10 while for amorphous materials is 3 (examples: the ε_r for Ta_2O_5 and Fe_2O_3 are 25 and 12.5, respectively. Note— larger values were reported for Fe_2O_3).
ε_o	Vacuum permittivity, also known as the permittivity of free space. 8.85×10^{-12} F/m.
η_{e^-/h^+}	Efficiency of photon absorption to charge excitation. Related to absorptance.
$\eta_{transport}$	Efficiency of charge transport to surface.
$\eta_{interface}$	Efficiency of interfacial charge transfer.
λ	Wavelength of light (nm).
Φ	Photon flux (photons/s).
ϕ_{bi}	Built-in potential: A term representing the value of bending of the valence and conduction bands downward at the junction.

Z. Chen et al., *Photoelectrochemical Water Splitting*,
SpringerBriefs in Energy, DOI: 10.1007/978-1-4614-8298-7,

σ	Absorption cross section for a particular electronic transition.
A	Absorbance: A mathematical quantity that is an approximation of absorption (not distinguishing among scattering, reflection, and transmission) and is defined by the Beer-Lambert Law: $A = -\log\left(\frac{I}{I_0}\right)$, where I is the intensity of the light transmitted through a sample and I_0 is the intensity of the light entering the sample.
$A_\%$	Absorptance: the fraction of incident light that is absorbed.
ABPE	Applied bias photon-to-current conversion efficiency: A term describing the photocurrent collected per incident photon, taking into account the external applied bias required relative to the water splitting voltage of 1.23 V.
AM 1.5 G	AM 1.5 G is a shorthand notation used by the PV community to denote the air-mass 1.5 global reference spectrum, estimated at 1000 W/m^2.
APCE	Absorbed photon-to-current conversion efficiency. A term describing the photocurrent collected per incident photon actually absorbed. It is the IPCE normalized to the IQE.
c	The concentration of the absorbing species. A term used in Beer's Law that is related to the amount of light absorbed for a given sample $A = \varepsilon cl$.
C	Capacitance: Total capacitance for a parallel-plate capacitor. Electric charge stored per electric potential. $C = \varepsilon A/d$, where A is the area of the plates and d is the distance between plates. The SI unit is the farad: 1 farad = coulomb/volt.
C_{dl}	Double layer capacitance: Excess charge in an electrode surface is compensated by a build-up of opposite-charged ions (Helmholtz layer), creating an electrical double layer. This layer is mathematically treated as a parallel plate capacitor. Typical values are on the order of tens of micro farads per cm^2.
CE	Counter electrode: In an electrochemical cell, the current flows between CE and WE.
C_{sc}	Space charge capacitance: In order to attain equilibrium between Fermi level and electrolyte redox, there will be a region within the material where a depletion of charges (and an electric field) will occur. The capacitance over this region is known as the space charge capacitance. Typical values are on the order of tens of nano farads per cm^2.

C_{ss}	Surface state capacitance. Surface states at a semiconductor-electrolyte interface allow for charge buildup via a capacitive effect.
e	The absolute value of the elementary electron charge $(1.602 \times 10^{-19}$ Coulombs).
e^-	Quasi-free electron. Refers to an electron free to move about within a bulk lattice.
$E(x)$	Electrical field as a function of distance, x (V/cm): The electrical field that results from the distribution of charges in the depletion layer at a certain distance into the semiconductor.
E_c	Lowest energy level of the conduction band (eV). Also referred to as the LUMO.
$E_{dark\text{-}onset}$	Dark current onset potential: The potential at which there is an onset of current under dark conditions.
E_F	Fermi level. An energy level that has 50 % probability of being occupied by an electron. The probability of occupancy decreases above the Fermi level (toward vacuum), and increases below the Fermi level. The Fermi level is near the middle of the band gap for an intrinsic semiconductor, near the conduction band for an n-type semiconductor, and near the valence band for a p-type semiconductor.
E_{fb}	Flat-band potential: The Fermi level on an electrochemical scale within a semiconductor electrode under no band bending conditions (no potential difference between the surface and the bulk energy bands).
$E_{F,solution}$	Fermi level in electrolyte.
E_g	An energy (eV) range devoid of electron states in a semiconductor: $E_g = LUMO - HOMO = E_c - E_v$.
$E_{h\nu}$	Photon energy (eV) $= h\nu$ (h is the Planck constant and ν is the photon frequency).
EIS	Electrochemical impedance spectroscopy: An alternating current technique that measures the impedance of an electrochemical system. The resulting spectra can be used to estimate the various components of an equivalent circuit that represent an electrochemical cell.
E_{max}	Maximum electric field at the interface, $x = 0$ (V/cm). See $E(x)$.
E_{onset}	Photocurrent onset potential: The potential at which there is an onset of photocurrent.

$E_{\text{onset,HER}}$ Onset potential for the HER.

$E_{\text{onset,OER}}$ Onset potential for the OER.

EQE   External quantum efficiency: This is same as IPCE.

E_{ref} Potential versus reference electrode.

E_{v} The highest energy level of the valence band (eV). Also the HOMO.

$E_{\text{WE–CE}}$ The potential between the working and counter electrodes. The absolute value is called the cell voltage.

f AC frequency (Hz).

F Faraday constant ($F = eN_{\text{A}}$) (9.648×10^4 C mol^{-1}): The magnitude of electric charge per mole of electrons.

Forward bias   The direction of applied bias toward majority band in the semiconductor relative to its OCP. It is anodic for p-type and cathodic for n-type semiconductor electrodes.

FTO   Fluorine-doped tin oxide, a transparent conductive oxide, typically used as a substrate for photoelectrodes.

FWHM   Full width at half maximum; width of the peak at half its maximum height.

ΔG Gibbs free energy difference. At standard temperature and pressure, this is denoted as ΔG_0.

h Planck's constant (6.626×10^{-34} J/s): A physical constant used to describe the magnitude of quanta in quantum mechanics.

h^+ Hole: An electron excited to the conduction band leaves behind a vacancy in the valence band. This hole represents the absence of an electron and has a positive charge of the same magnitude as an electron.

HER   Hydrogen evolution reaction. In acid, this half-reaction proceeds as $2H^+ + 2e^- \rightarrow H_2$. In base, this half-reaction proceeds as $2H_2O + 2e^- \rightarrow H_2 + 2OH^-$.

HOMO   Highest occupied molecular orbital. See E_{v}.

HOR   Hydrogen oxidation reaction. In acid, this half-reaction proceeds as $H_2 \rightarrow 2H^+ + 2e^-$. In base, this half-reaction proceeds as $H_2 + 2OH^- \rightarrow 2H_2O + 2e^-$.

I_0 Incident or input light intensity: A measure of the time averaged energy flux of light incident or input into the system. Similar to irradiance.

I	Current (A).
I_{dark}	Current under dark conditions (A).
$I_{\text{illuminated}}$	Current under illuminated conditions (A).
I_{ph}	Photocurrent ($I_{\text{illuminated}} - I_{\text{dark}}$).
IPCE	Incident photon-to-current efficiency. Also called an action spectrum: A measure of electron-hole pair collected (at a specific wavelength) per incident photon impinging on the photoelectrode surface. Also referred to as EQE.
IQE	Internal Quantum Efficiency: Same APCE.
Irradiance	The electromagnetic radiation that is incident on the surface ($\text{W m}^{-2}\,\text{nm}^{-1}$).
ITO	Indium Tin Oxide, a transparent conductive oxide used as a substrate for PEC electrodes.
I–V	A current (I) versus voltage (V) measurement. See LVS.
j	Current density (current/area).
j_{dark}	Current density under dark condition.
$j_{\text{E}^0=\text{HER}}$	Current density at the reversible potential for HER.
$j_{\text{E}^0=\text{OER}}$	Current density at the reversible potential for OER.
$j_{\text{illuminated}}$	Current density under illumination.
j_{ph}	Photocurrent density (current/area) ($J_{\text{illuminated}} - J_{\text{dark}}$).
j_{mono}	Photocurrent density using monochromatic light.
j_{sc}	Photocurrent density obtained under short circuit condition (or zero applied bias between the WE and CE).
j–V	Current density versus voltage.
k_{B}	Boltzmann constant (1.381×10^{-23} J K^{-1} or 8.617×10^{-5} eV K^{-1}).
l	Path length of light through the sample: a term used in Beer's Law and optical absorption measurements in order to determine the amount of light absorbed for a given sample.
l_{depth}	Optical penetration depth $= 1/\alpha$, where α is absorption coefficient which is a function of wavelength.
LSV	Linear Sweep Voltammetry: A voltammetric technique where current between working and counter electrode is measured while

linearly sweeping the potential of the working electrode. Also referred to as I–V or J–V measurements.

LUMO
Lowest Unoccupied Molecular Orbital. See E_c.

MBE
Molecular Beam Epitaxy: a standard ultra-high-vacuum vapor phase growth technique for producing single crystal films epitaxially on either homo- or hetero-single crystal substrates.

M–S
Mott–Schottky analysis: An electrochemical technique measuring impedance as a function of applied bias. The resulting data is analyzed to estimate the flatband potential, doping type and dopant density.

N_1
Population of states with electron energy level, E_1: a term used in UV–Vis spectroscopy in order to determine the amount of light absorbed for a given sample.

N_2
Population of states with electron energy level, E_2: a term used in UV–Vis spectroscopy in order to determine the amount of light absorbed for a given sample.

N_A
Acceptor doping density: Number of electrically active acceptors per volume ($\#/\mathrm{cm}^3$).

N_D
Donor doping density: Number of electrically active donors per volume ($\#/\mathrm{cm}^3$).

N_{dopant}
Dopant density of a semiconductor.

NHE
Normal hydrogen electrode. See SHE.

OCP
Open Circuit Potential: Measured potential difference between the WE and RE (immersed in electrolyte) under open circuit conditions. Also referred to as OCV.

OER
Oxygen Evolution Reaction. In acid, the half-reaction proceeds as $2\mathrm{H_2O} \rightarrow \mathrm{O_2} + 4\mathrm{H^+} + 4\mathrm{e^-}$. In base, the half-reaction proceeds as $2\mathrm{O_2} + 4\mathrm{H^+} + 4\mathrm{e^-} \rightarrow 4\mathrm{OH^-}$.

ORR
Oxygen Reduction Reaction: In acid, the half-reaction proceeds as $\mathrm{O_2} + 4\mathrm{H^+} + 4\mathrm{e^-} \rightarrow 2\mathrm{H_2O}$. In base, the half-reaction proceeds as $4\mathrm{OH^-} \rightarrow 2\mathrm{O_2} + 4\mathrm{H^+} + 4\mathrm{e^-}$.

PEC
Photo-Electro-Chemical or Photo-Electro-Chemistry: An electrochemical reaction performed utilizing light.

pH
$-\log[\mathrm{H^+}]$: a measure of the acidity or basicity of a solution.

P_{mono}
Calibrated light intensity (W/nm) as a function of wavelength.

P_{total}
Total integrated input power from the impinging illumination (W).

PV	Photovoltaic.
PVD	Physical vapor deposition: A class of techniques such as evaporation, sputtering, and ablation used to deposit thin films.
q	Charge of a particle.
Q	Bulk charge density: A term relating the amount of charge in a particular area. In p-type semiconductors this term is equal to qN_A and is generally measured in Coulombs/cm^3.
R	Resistance $(\Omega) = V/I$. R_s denotes the series resistance in a circuit.
RE	Reference Electrode: A nonpolarizable electrode that generates a stable and highly reproducible potential.
Reverse bias	The direction of applied bias toward the minority band in the semiconductor relative to OCP. It is cathodic for p-type and anodic for n-type semiconductor electrodes.
RHE	Reversible hydrogen electrode. The reversible potential for HER and HOR at a given condition.
SCE	Standard Calomel Electrode: A reference electrode based on the equilibrium reaction between elementary mercury and mercury(I) chloride.
SHE	Standard Hydrogen Electrode: The reversible potential for the hydrogen evolution and oxidation reactions at standard electrochemical conditions such that the activity of $H^+ = 1$. This potential is treated as 0 V on the electrochemical potential scale. Often used interchangeably with the normal hydrogen electrode (NHE).
STH	Solar-to-hydrogen efficiency. Photocurrent efficiency obtained under zero-bias conditions and using AM 1.5 G light spectrum.
t	Time (s).
T	Temperature (K).
$V(x)$	Potential as a function of distance, x, into the semiconductor (V) from the interface $V(x) = -(qN_A/(2\varepsilon_s))(w_d - x)^2 (0 \leq x \leq w_d)$
V_b	Applied bias (V) between the WE and CE.
V_{min}	The minimum voltage difference required between the WE and CE to drive water splitting.
V_{max}	Maximum potential at interface, $x = 0$ (V). See $V(x)$.
$V_{onset,HER}$	The potential difference between the onset of photocurrent for the HER versus the reversible potential for HER.

$V_{\text{onset,OER}}$ The potential difference between the onset of photocurrent for the OER versus the reversible potential for OER.

V_{ph} Photovoltage: the change in the OCP of the semiconductor in the dark versus under illumination, as measured versus RE.

w_{d} Depletion width; Width of a space region within a semiconductor associated with the depletion of mobile carriers due to equilibration between the semiconductor Fermi level and electrolyte redox level.

WE   Working Electrode.

Z Impedance. The component that is in-phase is referred to as the real impedance (Z') while the component that is out-of-phase is referred to as the imaginary impedance (Z'').

MIX
Papier aus verantwortungsvollen Quellen
Paper from responsible sources
FSC® C105338

If you have any concerns about our products,
you can contact us on
ProductSafety@springernature.com

In case Publisher is established outside the EU,
the EU authorized representative is:
**Springer Nature Customer Service Center GmbH
Europaplatz 3, 69115 Heidelberg, Germany**

Printed by Libri Plureos GmbH
in Hamburg, Germany